Environmental Policy Failure

The Australian Story

Kate Crowley
KJ Walker
Editors

Environmental Policy Failure: The Australian Story
1st edition, 3rd printing

Editors
Kate Crowley & KJ Walker

Cover designer
Christopher Besley, Besley Design.

ISBN: 978-0-7346-1140-6

Published in Australia by:
Tilde Publishing and Distribution
PO Box 72
Prahran VIC 3181 Australia
www.tup.net.au

TUP-EnvironPolFail-1e3p

Contents

About the editors

Associate Professor Kate Crowley is Head, School of Government and lectures in public and environmental policy at the University of Tasmania, Hobart, Australia. Kate was introduced to environmental policy in a practical sense at Monash University as a member and then President of the Monash Bushwalking Club. She has represented Australia in cross country skiing as a competitor and instructor, and in her academic career is widely published on environmental governance, green politics and public policy. She has previously edited, with Ken Walker, *Australian Environmental Policy: Studies in Decline and Devolution* and she consults at the environmental policy interface, for example, by writing the *Framework for Reducing Tasmanian Government Emissions.* Currently Kate Chairs the Tasmanian Climate Action Council, which is an independent statutory authority that reports to Climate Change Minister and Tasmanian Greens Leader Nick McKim, the first Green minister in Australia.

Dr K J Walker is an independent researcher and a pioneer of the study of environmental politics in Australia. His books include *The Political Ecology of Environmental Policy: An Australian Introduction, Australian Environmental Policy* (editor) and *Australian Environmental Policy: Studies in Decline and Devolution* (co-editor). He has written many influential papers including 'Ecological Limits and Marxian Thought', *Politics*, Vol. XIV, No. 1, May, 1979, pp. 29-46, and 'The State in Environmental Management: the Ecological Dimension', *Political Studies,* Vol. XXXVII, March 1989, pp. 25-39. Ken has taught at Melbourne, Monash, Flinders, Adelaide, Case Western Reserve, Cleveland State, Griffith, Queensland and Macquarie universities. From the mid-1970s, he developed and taught ground-breaking interdisciplinary courses in environmental policy and politics. He writes mainly in the fields of environmental policy and environmental political theory, with subsidiary interests in technology, technology transfer and history in its social context.

About the contributors

Dr Michael Buxton is Professor, Environment and Planning in the School of Global Studies, Social Science & Planning, RMIT University. Before joining RMIT University, Michael held senior management positions with Victorian Government planning and environmental agencies, and has contributed extensively to national and state environmental and planning policy. He has held many government positions including the position of chairperson of the Premier's Green Wedge Working Party that advised the Victorian government on the adoption of a Melbourne green belt and legislated urban growth boundary. His recent research interests include major research projects into Australian peri-urban issues funded by Land and Water Australia and the Victorian government.

Associate Professor Geoff Cockfield is Deputy Dean of the Faculty of Business and a member of the Australian Centre for Sustainable Catchments (ACSC) at the University of Southern Queensland, Toowoomba, Australia. Geoff's research focuses on agri-environment policy, farm forestry policy and regional development policy. His funded research projects include the location and extent of agricultural production under climate change, the socio-economic impacts of forest vulnerability under climate change, and the socio-economics of sustainable grazing systems.

Daniel Connell is a Research Fellow in the Crawford School of Economics and Government at the Australian National University in Canberra. Daniel teaches courses in communications and environmental policy and the governance of trans-boundary rivers. His research focuses on large rivers in the federal systems of Australia, South Africa, the United States, Mexico, the European Union (Spain), India, China and Brazil. He also contributes to the ongoing debate about the future of the Murray Darling Basin – the subject of his book *Water Politics in the Murray-Darling Basin* published in 2007.

Dr Aysin Dedekorkut-Howes is a Senior Lecturer at Griffith University School of Environment's Urban and Environmental Planning Program. She is a member of the Griffith Climate Change Response Program and co-convenor of the Gold Coast Area Research Team specialty subgroup of the Urban Research Program. Her research interests include climate change adaptation; multi-level governance for regional level planning and growth management; and urbanisation and development in subtropical cities.

Associate Professor Mark Diesendorf is Deputy Director of the Institute of Environmental Studies at University of New South Wales where he teaches, researches and consults in the interdisciplinary fields of sustainable energy, energy policy, sustainable urban transport, theory of sustainability, ecological economics, and practical processes by which government, business and other organisations can achieve ecologically sustainable and socially just development. He is co-editor of the interdisciplinary book *Human Ecology, Human Economy: Ideas for an Ecologically Sustainable Future* (1997) and author of *Greenhouse Solutions with Sustainable Energy* (2007). His latest book, *Climate Action: A Campaign Manual for Greenhouse Solutions,* was published in 2009.

Dr Michael Howes is a Senior Lecturer at the Griffith School of Environment, a researcher with the Urban Research Program and a member of the Griffith Climate Change Response Program. He specialises in environmental policy with a particular focus on climate change, sustainable development and ecological modernisation. Michael is the author of the book *Politics and the Environment: Risk and the role of government and industry* published internationally by Earthscan.

Professor Ian Lowe AO is Professor of Science, Technology and Society at Griffith University, Adjunct Professor at Sunshine Coast University, and Honorary Research Fellow at the University of Adelaide. He is President of the Australian Conservation Foundation, and in 2001 was made an Officer of the Order of Australia for services to science, technology and environmental studies. He was been awarded a Centenary Medal for contributions to environmental science, and the Eureka Prize for promotion of science. His work has been recognised by the Prime Minister's Environment Award for Outstanding Individual Achievement, and the Queensland Premier's Millennium Award for Excellence in Science.

Associate Professor David Mercer is Program Director of the Masters in Social Science (International Urban & Environmental Management) and Environment & Planning degrees in the School of Global Studies, Social Science & Planning, at RMIT University, Melbourne. He has degrees in Geography from Cambridge and Monash universities, and in 1999 was elected a Fellow of the Environment Institute of Australia and New Zealand (EIANZ) in recognition of his services to environmental research, consultancy and teaching. From 2002 to 2009 he served as a Council Member on the Victorian Environment Assessment Council (VEAC). He has served on the editorial boards of several journals - most recently the *Australian Journal of Natural Resources and Policy*, the *Open Environmental Journal* and the *Annals of Leisure Research.*

Mark O'Connor is a poet and environmentalist, the author of *This Tired Brown Land*, co-author of *Overloading Australia: How Governments and Media Dither and Deny on Population* (with William Lines), and an editor of the textbook *Protected Area Management*. He has published more than a dozen other books and received several awards. With Judith Wright, he founded Writers for an Ecologically Sustainable Population, and served for a decade as National Vice-President of Sustainable Population Australia. He has taught at James Cook University, the University of Aarhus and the Australian National University, and has been the ANU's H C Coombs Fellow and the Museum of Victoria's Thomas Ramsay Science and Humanities Fellow.

Joan Staples is Visiting Fellow in the Faculty of Law at the University of NSW where she teaches a course on Strategic Public Advocacy for Civil Society. Her research interests and publications have focussed on the role of civil society and, in particular, civil society's place in the democratic process. Her academic work references a career in policy and advocacy for national non-government organisations, including as National Liaison Officer of the Australian Conservation Foundation during the Hawke Government in the 1980s. Her advocacy work and board appointments have been with national NGOs relating to environment, Indigenous affairs, consumers, international development, women's affairs and social services.

Associate Professor Geoffrey Wescott is Associate Professor of Environment in the School of Life and Environmental Sciences at Deakin University in Melbourne, and Vice President of the Australian Coastal Society. He is a former chair of the Marine and Coastal Community Network and member of the National Oceans Advisory Group. He is also a former member of the Victorian Coastal Council and Parks Victoria Board. Recent books include *Back to Basics – Breakthrough Proposals for the Australian Environment* and *Rocky Shores of South Eastern Australia*. Geoff has used the coastal environment in Victoria as a long-term case study of the development, implementation and review of coastal policy based on the international concept of integrated coastal zone management.

Foreword

It should be no surprise that the best scholars in Australia conclude that environmental policy has failed. Nearly forty years after the first national environmental legislation, almost all the significant environmental indicators are still in decline. Three independent reports on the state of the environment have documented the loss of biodiversity, the decline of inland rivers, pressures on the coastal zone, degradation of productive land, and spiralling greenhouse gas emissions. In this important book, specific studies of such issues as energy and climate change, the Murray-Darling Basin, the coastal zone, water policy, natural resource management and population lead to the same conclusion: all elected governments – Commonwealth and state, ALP and Coalition – have as their highest priority economic growth through the exploitation of natural resources. Rather than seeing their duty as protecting social integrity and the environment, which would require assessing proposals for economic development against that higher responsibility, their obsession with short-term economic gain overrides other considerations: not just the integrity of natural systems but also social inclusion, the rights of Indigenous people and participative democracy are ruthlessly sacrificed for the sake of economic growth. The policy failures are therefore no accident, but the inevitable consequence of short-term thinking and the absurd emphasis on the economy, despite all the evidence of accumulating problems.

After reading this book you will be much better informed about the failure of environmental policy. I hope you will also be determined to attack the root cause of that failure, our dominant political culture and the values it embodies.

Ian Lowe
President
Australian Conservation Foundation
October 2011

Preface

This book follows two previous successful readers: *Australian Environmental Policy* (Walker 1992) and *Australian Environmental Policy: Studies in Decline and Devolution* (Walker & Crowley 1999). It was conceived during the dark years, especially for environmental policy, of the Howard Coalition federal government (1996-2007). These years saw federal hostility to the science of global warming, and to environmentalism and green politics in general, coupled with enhanced enthusiasm for the 'quarry Australia' vision of resource exploitation. Although climate change inaction and friction with environmental groups was nothing new, there was a particular gloom about the tenor of the conservative government that was deepened by affairs extraneous to this book's central focus. Our earlier studies into the decline and devolution of environmental policy responsibility, coupled with these more recent developments and the ever deteriorating state of the environment, suggested that a more critical treatment was required.

We embarked upon this book with an interest in recording environmental policy dilemmas over recent decades, only to be struck by the high degree of recurrent policy inadequacy, neglect and failure that we encountered. We ranged across a broad set of environmental issues of national significance, seeking to identify lessons to be learned for effective policy-making and politics. Coincidentally, we set about this task at a time when major natural disasters were on the rise – with enduring drought but also with horrendous bushfires, massive flooding and severe tropical cyclones. Were these 'acts of God' or were they the consequences of policy failure? Whether they were or were not caused or aggravated by human activity can only be conjectured in the absence of reliable data and expert interpretation. But such a year of natural disasters, as seen in Australia in 2010, at least reminds us of our place in nature and of our great responsibilities as custodians of it.

The Rudd Labor government's landslide election in 2007 and subsequent ratification of the Kyoto Protocol brought hope of meaningful climate policy action. But Labor offered no clear agenda for a broad recovery in environmental policy, and any heightened public expectations following the defeat of the conservatives were dashed. There is a line of historical continuity in environmental policy failure that is unbroken by changes of government. And so it was that Labor's policy promise faltered in a blizzard of hesitant mediocrity. Even on climate change, for which Labor had an unchallenged mandate for action, they produced a watered down and ineffective emissions trading scheme that pleased no one. It was obvious to all that many of the forces that had driven climate policy, in particular under the Howard government, were still operative and still continuing to frustrate effective policy. Indeed, in the search for explanations of policy failure in the environmental field, we were struck in passing by the slow implosion of two-party politics and the rise of the Greens, reflecting a public concern for the environment that has persisted for decades.

Our focus on policy failure shines a light on the incapacity of mainstream politics to effect change. This is concerning because policy success requires political commitment. Our message is very clear: there are ready policy solutions to Australia's environmental problems that could be adopted immediately given sufficient political will. We are not the first to suggest this. Our findings support those of earlier studies: Ian Lowe's *A Big Fix: Radical Solutions for Australia's Environmental Crisis* for example; or David Lindenmayer's *On Borrowed Time: Australia's Environmental Crisis and What to Do About It;* or Stephen Dovers and Su Wild River's *Managing Australia's Environment.* At the global level, the seriousness of environmental policy failure is emphasised by the 2005 Millennium Ecosystem Assessment *Living Beyond Our Means* which calls for the urgent halting of environmental decline, the valuing of nature, and the mainstreaming of protective measures. Australia must commit to just such effort before it is too late.

Kate Crowley and KJ Walker
October 2011

Table of tables

Table of figures

Acronyms

AAP	Australian Associated Press
ABARE	Australian Bureau of Resource Economics
ABC	Australian Broadcasting Corporation
ABS	Australian Bureau of Statistics
ACS	Australian Coastal Society
ACF	Australian Conservation Foundation
ACTU	Australian Council of Trade Unions
ACS	Australian Coastal Strategy
AOP	Australia's Oceans Policy
ACRE	Australian Centre for Renewable Energy
AGPS	Australian Government Publishing Service
ALP	Australian Labor Party
CEWH	Commonwealth Environmental Water Holder
CCTV	closed-circuit television camera
CCMP	Climate Change Management Plan
CCS	carbon capture and storage
CFCs	chlorinated fluorocarbons
CO_2	carbon dioxide
COAG	Council of Australian Governments
CPRS	Carbon Pollution Reduction Scheme (Rudd Government) or Continue Polluting Regardless Scheme (Australian Greens)
CRC	Co-operative Research Centre
CSIRO	Commonwealth Scientific and Industrial Research Organisation
DEFRA	Department for Environment, Food and Rural Affairs
DERM	Department of Environment and Resource Management (Queensland)
DCC	Department of Climate Change
DIP	Department of Infrastructure and Planning (Queensland)
DRET	Department of Resources, Energy and Tourism

EE	Energy Efficiency (more precisely, efficient energy use)
EEZ	Exclusive Economic Zone
ENGOs	Environmental Non Government Organisations
EPBC	Environment Protection and Biodiversity Conservation
ERET	Extended Renewable Energy Target
ESP	Environmental Stewardship Program
EREC	European Renewable Energy Council
ETS	emissions trading scheme
EU	European Union
FOE	Friends of the Earth
GCCC	Gold Coast City Council
GFC	global financial crisis
GGAP	Greenhouse Gas Abatement Program
GL	gigalitres
HORSCCCWEA	House of Representatives Standing Committee on Climate Change, Water, Environment and the Arts
ICM	Integrated Catchment Management
ICZM	Integrated Coastal Zone Management
IPCC	International Panel on Climate Change
LETDF	Low Emission Technology Demonstration Fund
MCCN	Marine and Coastal Community Network
MBI	Market-based instrument
MDB	Murray-Darling Basin
MDBA	Murray Darling Basin Authority
MDBC	Murray–Darling Basin Commission
Migrant	means 'immigrant' in Australian English
ML	megalitres
MPA	marine protected area
MRET	Mandatory Renewable Energy Target (Howard government scheme)
NAP	National Action Plan for Salinity and Water Quality
NCCARF	National Climate Change Adaptation Research Facility
NELA	National Environmental Law Association
NEPC	National Environmental Protection Council

NETT	National Emissions Trading Taskforce
NGOs	Non Government Organisations
NHT	Natural Heritage Trust
NLP	National Landcare Program
NOAG	National Oceans Advisory Group
NOO	National Oceans Office
NOM	Net overseas migration
NOMB	National Oceans Ministerial Board
NRM	Natural resource management
NRMMC	Natural Resource Management Ministerial Council
NRS	National Reserve System
NSCP	National Soil Conservation Program
NSTF	National Sea Change Task Force
NWC	National Water Commission
NWI	National Water Initiative
OECD	Organisation for Economic Cooperation and Development
PV	Photovoltaic
RAC	Resource Assessment Commission
R&D	research & Development
RD&D	research, development & demonstration
RE	renewable energy
REC	Renewable Energy Certificate
REDI	Renewable Energy Development Initiative
REDP	Renewable Energy Development Program
REEF	Renewable Energy Equity Fund
RElec	renewable electricity
RET	Renewable Energy Target (Rudd government's expanded scheme)
RFA	Regional Forest Agreement
RMC	River Murray Commission
RMP	Regional Marine Plan
RRPGP	Remote Renewable Power Generation Program
SDL	sustainable diversion limits
SEQ	South East Queensland

SMH	Sydney Morning Herald
SPA	Sustainable Population Australia
SPPA	Stable Population Party of Australia
TWS	The Wilderness Society
UN	United Nations
UNEP	United Nations Environment Program
VNPA	Victorian National Parks Association
WFD	Water Framework Directive
WUI	wildland-urban interface
WWF	World Wildlife Fund

Chapter 1

Introduction[1]

KJ WALKER AND KATE CROWLEY

Introduction

A central theme of this book is that Australian policy-makers have persistently ignored the limitations of the Australian environment; physical, climatic and ecological. The consequences have been dire, affecting every aspect of the nation's ecological support system: over-allocation of water from the Murray-Darling river system; peri-urban development in areas of high fire danger; destruction of native forests by logging and wood-chipping; extensive land clearing for agriculture, grazing and settlement; conversion of fertile land for plantations, roads and housing; and atmospheric, inland aquatic, marine and estuarine pollution. And population growth on the fragile Australian continent, where lifestyles are amongst the most consumptive and wasteful in the world and where only the margins are habitable and arable, continues to deliver unsustainable ecological impacts.

These are longstanding problems, some as old as European settlement. But warnings of ecological fragility, over-settlement impacts and environmental policy failures also date back almost as long. They have been conveyed loudly and strongly for at least fifty years, for example in: *Last of the Lands* (Webb *et al.* 1969); *A Dirty Story: Pollution in Australia* (Whitelock 1971); *The Politics of Finding Out* (Dempsey 1974); *Cities for Sale* (Sandercock 1976); *Environmental Management in Australia* (Powell 1978); *The Australian Environment: Twelve Controversial Issues* (Gilpin 1980); *Spoils and Spoilers* (Bolton 1981); *Quarry Australia* (Birrell & Stanley 1982); *Saving A Continent: Towards a Sustainable Future* (Smith 1994); *The Future Eaters* (Flannery 1994); *Managing Australia's Environment* (Dovers and Wild River 2003); *A Big Fix: Radical Solutions for Australia's Environmental Crisis* (Lowe 2005); *On Borrowed Time: Australia's Environmental Crisis and What We Must Do About It* (Lindenmayer 2007); *High and Dry: John Howard, Climate Change and the Selling of*

[1] Aspects of this chapter draw on previous work by both authors.

Australia's Future (Pearse 2007); and *Scorcher: The Dirty Politics of Climate Change* (Hamilton 2007).

Thirty years ago Birrell and Stanley (1982) identified a 'quarry Australia' development mentality as driving environmental destruction. This theme endures today, as shown by the work of Mercer and Marden (2006) who observe a 'consistent resistance to a strong environmental policy agenda' (p. 183) in this country. Pearse (2009) also finds that the 'quarry vision' persists and that it values Australia's mineral and energy assets, essentially coal, as its greatest riches, in turn distorting our national politics and engendering a shadowy world of carbon lobbyists with very narrow agendas.

Before revisiting the question of whether environmental policy succeeds or fails in Australia, it is worth restating the links between the fragile nature of the continent's ecology, its political economy and the enduring threats to its environment. This is because describing environmental policy in Australia is not simply a question of identifying policy processes, institutions and legislation to see how well they work, or conversely how badly they fail. Analysis must also consider critical contextual issues such as policy constraints, missed opportunities and failed initiatives, non-decisions as well as decisions, and issues that have been ignored and repressed as much as those that have been embraced. It is timely to do so, given recent environmental crises in Australia, the long-term drought, and the more recent catastrophic bushfires, cyclones and floods.

Facing environmental crisis

In February 2009, the southern state of Victoria experienced the worst bushfire disaster in Australian history on what has become known as Black Saturday, when 173 people lost their lives, 414 were injured and over 2,000 homes were lost. Tinder-dry as a consequence of twelve years of drought, and in record temperatures of 45 degrees Celsius, the fires ran wild, fanned by high winds. Several substantial communities, including Kinglake and Marysville, were burnt out. Other major communities, such as Yea, were threatened, as were outlying metropolitan suburbs of Melbourne such as Belgrave in the lower Dandenong Ranges.

While commentators were apportioning blame and forestry experts were promoting their pet nostrums, state and federal governments and local communities were pledging to rebuild the towns in their entirety, against the best advice. But climatologists and meteorologists were also pointing to the extraordinary and unprecedented conditions that had resulted in the disaster. The combination of elevated temperatures, high winds and an abundance of drought-baked fuel had never been so intense before, eclipsing even the conditions preceding the devastating 1983 Ash Wednesday bush fires.

The heat at the time and the decade-long drought were attributed to a long-term global warming trend. While it is difficult to link any one event to global warming, the trend predicted by climate science is for more intense weather patterns (Kelly 2011; Lauder 2010). Severe fluctuations or disruptions in weather patterns are also expected, and in recent years have seemed perceptible in Australia. The Pacific's El Niño/La Niña circulation patterns have been irregular for some years; a similar pattern in the Indian Ocean, on which the Indian monsoon depends, has also been showing signs of disruption. Meanwhile, the long-term drought in southern Australia was being attributed to a southward shift in the regular progression of storms across the region, warming the climate and starving it of formerly regular rainfall. The effects were devastating but there was worse to come.

While attention had been paid to localised impacts, for example the impact of long-term drought on agriculture and river systems, less attention had been paid to the growing evidence that accelerating climate change is most likely already upon us. This was certainly speculated upon over the 2011 summer, which brought the destructive force of cyclone Yasi, a category five tropical cyclone, to bear on the northern Queensland coast. Yasi hit with wind gusts of 290 kilometres per hour and storm surges of seven metres, devastating not only to communities, agriculture, industry and the environment, but, in economic terms, inflicting costs that will take billions to cover.

Cyclone Yasi hit Queensland in early February 2011 in the wake of a series of devastating floods that had begun in Queensland in December the year before and which, at their deadliest, flash-flooded Toowoomba, almost wiped out Grantham in the Lockyer Valley, and flooded the centre of Ipswich. The seven-metre high flood waters crossed country on waterlogged soils, forming a terrifying 'inland tsunami', before moving on to flood the Brisbane River catchment, the Brisbane central business district and surrounding suburbs to a height of four metres. Thirty-seven people lost their lives, with some bodies swept eighty kilometres and fears that others would never be found. The after-effects were felt across the inland to central Australia before sweeping downwards into the Victoria with flooding rain afflicting regions which had been experiencing flooding since late 2010. The 2011 Victorian floods were declared by the Bureau of Meteorology to be the worst in history in terms of rainfall totals, depth of water and the number of places affected (Cauchi 2011).

Predictions are that the low-lying Australian continent will be one of the hardest hit by the effects of global warming and climate change, including stronger exposure to counter-adaptive trends such as cyclones and storm surges, and more frequent and severe major droughts (Pittock 2009, pp. 284-6). But is the Australian environment robust enough to withstand this? Indeed, how resilient is it in the face of growing ecological crisis, whether human induced or natural?

The wide brown land

Australia is unique in the sense that it is the world's driest inhabited continent, and so it is atypically vulnerable to the expected reductions in rainfall across populated regions and to the enhanced stress on water resources that global warming and climate change will inevitably produce over the long term. Flannery (1994) describes a unique, fragile and highly interconnected ecology unlike that of any other continent. Its fragility is evident in its vulnerability to the disruption of its ecology and productive systems and to the intensified impacts of its society, which still predominantly employs ill-suited Eurocentric agricultural practices but also has largely settled in concentrated fashion in just two per cent of the landmass along the eastern/southeastern coastline.

The Australian continent presents deep contradictions for the formulation of environmental policy. For example, it is the world's most mega-biodiverse continent, but it supports this biodiversity with the world's most nutrient-poor soils. It is the country's stores of biodiversity that supply its soils with nutrients, and so land, woodland and forest clearing are each key threatening processes, whilst programs to conserve biodiversity are key mitigating and protective measures.

With less than ten per cent of its soils arable, Australia's meagre rainfall, fragile soils and limited fertile land have always meant that even 'normal' fluctuations could disturb not just its environment but also its economy, especially those industries dependent on ecological factors: agriculture, pastoralism and agrarian endeavours. As most of the continent is marginal for agriculture and pastoralism, expansion along European lines has lagged since settlement, but its promotion, a central issue in politics, remains intimately tied to an unsustainable exploitation. Indeed, Australia is the most dependent of the OECD nations upon resource exploitation, with agriculture and mining accounting for over 61 per cent of export earnings derived from trade in commodities (OECD 2007).

Australia's extraordinary richness of biodiversity has resulted in the world's highest number of World Heritage listings for natural values alone. Two dozen areas are listed, including the tourist meccas of Kakadu, the Great Barrier Reef and the Tasmanian wilderness. Australia's unique fauna and flora are also of exceptional scientific interest and embody important aesthetic values (Smith 1990; Boyden Dovers & Shirlow 1989). But European settlement has accelerated the rates both of exploitation and of ecological damage. There are direct, immediate and large-scale threats to endemic flora and fauna: Australia's record for mammal species extinctions, for example, is the worst for any country (State of the Environment Advisory Council 1996, pp. 4-39).

Australia's settlement involved the transfer of European crops, livestock and techniques (Crosby 1986). Sparsely distributed and overmatched, much of the indigenous population was displaced, some of it exterminated outright, losing nearly all their land in the more hospitable, temperate regions. But in the arid centre and the tropics where European agricultural and pastoral technology was less effective, the indigenous population survives. Remnant natural areas and forests are also the areas of greatest conflict over conservation, since native fauna and flora have been devastated in the more settled areas (Bolton 1992; Lines 1991; Marshall 1966).

State of the Environment (SOE) reporting clarifies Australia's truly global responsibility for its biodiversity, with its conservation attempted through programs such as land health, freshwater health, vegetation conservation or marine protection. Threatening processes remain acute, however. Land clearing for urban development and agriculture continue to deplete and endanger native species and ecosystems, while many native species are killed every year on the roads which continue to dissect the remaining wildlife corridors. SOE reporting has confirmed that salinity, caused by land clearing and irrigation, and other pressures on water quality threaten not only plants, animals and ecosystems but also the future of agricultural and other industries in many parts of Australia, and even urban water supplies in some cities (Commonwealth 2002).

Yet another contradiction is that, despite Australia's vast empty, inhospitable desert spaces, it is also one of the most urbanised societies in the world, with serious urban pollution problems, affecting air, water and the disposal of toxic wastes. With its fertile coastal and inner coastal zones increasingly heavily populated, and with pressure on resources, especially water, steadily increasing, stress has become more and more apparent. The pressure of human activity and urban development takes a heavy toll on marine environments. Pollution is the most serious problem, with most marine pollution caused by land-based activities and processes, including soil erosion, fertiliser use, intensive animal production, sewage and urban industrial discharges. And it is widely accepted now that water shortages are a serious threat to Australia's economy and its way of life.

In sum, environmental policy issues in Australia differ greatly from those of European nations. Wilderness, the conservation of threatened species both floral and faunal, and issues affecting rural industries gain more emphasis while 'urban' and 'industrial' issues, notably pollution and waste management, although of critical importance, are very often less stressed. Acute public awareness of important, extensive, relatively undisturbed wilderness and unique flora and fauna has furthermore engendered conflict between developers and conservationists. Ingrained patterns of thought and behaviour are highly persistent in Australia, in particular in the post-settlement 'quarry Australia' resource exploitation paradigm which so influences decision makers. There is very little appreciation in this paradigm of the fragility, complexity and unique challenge of ecology.

The problem of ecology

Explanations of environmental policy success or failure must wrestle not only with the fragile nature of the Australian environment, but also with the problem of ecology. Ecology differs from other foci of political interest precisely because it is complex and fundamental, binding all living things together by a common dependence. Ecosystems are distinctive, place-based patterns of interactions that have evolved over millennia, to be delicately balanced in dynamic equilibrium. Once disrupted, by land clearing or clear-felling for example, ecosystems can lose their self-regulating, self-adjusting properties. Given the trans-boundary global connectivity of ecosystems, specific actions may impact at times or places quite remote from them. The interconnectivity of ecosystems can produce unexpected synergistic impacts, whereby substances and processes in the environment may, and frequently do, interact with each other in unpredictable, complex ways.

Risk projections that inform policy may not account for synergistic ecological impacts, calculating risk in isolation from other potentially related risks, and so they may fail to anticipate or greatly underestimate aggravated impacts. Moreover, ecological systems have thresholds or tipping points, below which adverse impact may not be detected but above which catastrophe, such as a blue-green algae outbreak on a major inland river, may suddenly occur. Having an awareness of the complexity of the problem of ecology is therefore a critical first principle for ensuring environmental policy success. Similarly, there is a need for policy-makers to appreciate uncertainty. Scientific knowledge is never final: it may be authoritative as the best and most broadly accepted knowledge available, but it is constantly reinterpreted and improved.

Moving from ecology to policy, we find that the social effects of environmental problems are extensive and far-reaching: often they too interact in novel ways. Many civilisations have collapsed, after all, from irreversible ecological problems that are all too familiar today, such as declining soil fertility, salination of irrigation systems and collapsing rural productivity. Diamond (2005) lists the twelve key environmental problems facing humankind today, the first eight of which have historically contributed to the collapse of past societies and most of which afflict contemporary Australia. These eight are: deforestation and habitat destruction; soil problems (erosion, salination, soil fertility loss); water management problems; overhunting; overfishing; effects of introduced species on native species; over-population; and the increased per capita impact of people.

Modern industrial societies are living on ecological capital to a dangerous degree, with abundant evidence of both ecological and policy failure rather than of the policy *learning* which should follow a history of collapse. In part this is a consequence of the failure to treat environmental concerns as social manifestations of underlying ecological malfunctions, and the failure to pay heed to the science that could inform policy. The result is often displacement, rather than the resolution, of an environmental problem. Many features of the contemporary world environmental crisis — excessive population, shortages of renewable and non-renewable resources, decline in the number of many species and the extinction of others, and the destruction of soil fertility, waterways and the atmosphere — are, in whole or part, consequences of displacement.

Part of the problem of enhancing ecological stewardship to remedy such crises is that it requires a rethink of industrial imperatives, material values, our political economy and our relations with the non-human world. Again, this is why environmental policy formation is a novel challenge. It must take notice of concepts such as the intrinsic value of nature, sustainability, and intergenerational equity that policy-makers must untangle and accommodate but that run counter to what is described as the dominant economic paradigm. On the other hand, the quest for greener policy has inspired new horizons, new economics and industries that are clean and green, new philosophy and new politics. But in Australia there are clear limits to how far a nation-state, economically reliant upon natural resource exploitation, is prepared to go towards adopting a protective ethic of care towards its environment.

Past practice is littered with policy failures. Externalities are not factored into the cost of production. Technological fixes may worsen problems. Once an ecosystem is damaged, sometimes in ways that are barely perceptible, there may be a risk of irreversibility. Given this risk, the 'precautionary principle' has long been advocated as a means of guarding against the ecological damage that may result from our ignorance and uncertainty. Precaution observes the central message of ecology, which is holism, connectivity and complexity.

Elements of policy success

To become more effective, environmental policy-making must include: the principles of holism, connectivity and complexity; an awareness of ecological thresholds or tipping points; and precaution in the face of uncertainty, ignorance and irreversibility. The focus upon ecology is a first principle when considering policy success. As Kraft and Furlong (2007, p. 348) observe, 'the true measurement of (environmental policy) success or failure is the status of ecosystem functioning or health', although they note that there is little agreement about what constitutes ecosystem health.

Beyond the first principle of ecosystem health, potential definitions of environmental policy success and, conversely, indicators of policy failure vary enormously in both theory and empirical analysis. Marsh and McConnell (2010) find that claims of policy success are rarely justified in a systematic way. They propose a simple framework of analysis with three dimensions: process success; programmatic success; and political success. But they also consider the issues of: success for whom; variations across time, space and culture; and methodology. Their starting point is that policy success means much more than simply meeting policy objectives or producing 'better' policy; it is also more broadly about puzzle solving, democratic dialogue and critique (p. 565-6).

The three dimensions vary from the narrow to the broad. First, *process*: was the policy legitimate (i.e. produced through due process, procedures and democratic values, deliberation and accountability)? Is it politically sustainable/well supported? Is it innovative, or does it draw on experience elsewhere? Secondly, *programmatic:* is the policy operational (implemented); did it achieve its aims; was it efficiently resourced; and in terms of actors/interests, who did it benefit? And finally the *political* dimension: is the policy politically popular? It is easy to see that analysis of policy success that is limited to process may draw very different conclusions from broader programmatic/political analysis.

Janicke and Weidner (1995) have given some thought to the definition of environmental policy success. Essentially, they describe a multi-dimensional set of criteria for evaluating the ecological improvement of policy efforts in a comparative context and describe success as the best or nearly best achievement. Their criteria take into account *character of the problem:* is it hard or easy to solve;

lasting effect: is the effect short, medium, long term, and does it solve the problem; *bonus effect:* is the ecological improvement related to the policy or is it a side-effect of other measures/actions; *problem-shifting:* was the ecological improvement achieved by shifting the problem elsewhere; *innovation:* was the improvement just an additive measure or was there a new/changed process; *efficiency:* has there been a net positive economic effect; *political acceptance*: is the improvement broadly accepted and so likely to last? (Janicke & Weidner 1995, pp. 15-16).

We can also look to the findings of environmental policy analysts for clues to the nature of environmental policy success/failure. Ison and Collins (2008, p. 15), for example, suggest that it is the failure to treat the management of the Murray-Darling river basin system as a 'wicked' problem that constrains effective action. If it were treated as wicked, then policy-makers would *abandon certainty:* and build an understanding of complexity; *invest in learning:* and use learning to build and support the capability to enact policy; *reflect on capacity*: by exploring ways to understand and reduce institutional complexity; *adopt holism*: and think and act holistically, or face unintended consequences. Within modern liberal democracies we tend to deal with environmental policy in a fragmented, silo-based fashion, whilst these authors advocate the opposite–namely that systems-based thinking and action is the key to policy success.

Finally, Dovers (2008) identifies issues that he believes need to be addressed in terms of improving environmental policy and achieving meaningful institutional reforms, which could be seen as ingredients for policy success. These are: 1) factor in the long term; 2) integrate the environment, society and the economy in policy; 3) ensure sustainability and its aims are clearly codified in statute law; 4) support international sustainability efforts; 5) adopt innovative policy processes; 6) involve the community; 7) decouple growth and environmental degradation; 8) refashion the statutory framework; 9) attend to overlooked sectors; and 10) establish a bi-partisan agenda.

About this book

There are various ways of thinking about environmental policy success and determining the nature and extent of policy failure. The chapters which follow illustrate many of these approaches. Ecosystem functioning or health is a primary consideration, but so too is political economy and the environmental consequences of Australia's reliance upon its natural resource base. Despite Australia's unique environment, its heavy reliance upon its exploitation is a constraining factor on environmental policy-making, and so context is therefore a critical dimension. Policy context provides scope for recognising the complex and challenging nature of ecological concern and environmental policy demands, as well as principles such as holism, uncertainty and precaution, which are key elements of policy success or failure. By adding policy context to Marsh and McConnell's three frames for considering policy success, we find that the range of issues considered above will broadly fit under four analytic criteria: policy context; process success; programmatic success; and political success. Much of

Australian environmental policy-making is highly politicised, fuelled by protest over 'places worth keeping' and systemic development pressures that are wholeheartedly supported by many politicians. And so, political success has particular significance as a criterion of analysis.

It is not possible for this book to comprehensively cover all issues in its quest to explain the multiple environmental policy failures in Australia that are now linked to so many endemic problems and ecological crises. Nor is it possible to cover from multiple perspectives each issue we have chosen. Our focus is not upon environmental science, management, economics, or even politics, for example, but upon policy. Our argument is that policy failure is longstanding and deeply rooted in over 200 years of Eurocentric thinking. We review some of the most critically challenging environmental issues of our time: climate change, energy policy, natural resource management, the Murray-Darling, marine and coastal policy, water policy, catastrophic bushfire, over-population, and the role of dissent.

This book brings together essays from prominent researchers in the field. They draw upon scholarly research, but in a broad policy context and in a manner that examines the relevance of the issues to the broader community. Each chapter ranges across the policy legacy of past governments, crucial political and economic issues, a consideration of the effectiveness of the policies adopted, and future implications for policy in each arena. Although there have been many books over the years looking at environmental issues, very few have had a concerted policy focus, or a policy focus within a political economy context. Two of the most recent useful texts in terms of policy analysis are from Dovers and Wild River (2003) and Thomas (2007), but even these are beginning to date.

We see the development and implementation of effective environmental policy as one of the most urgent, yet complex, tasks facing government today, and yet we are struck by the failed policy learning around the country. Hence, our focus is upon long-term policy failure and the need to interrogate it.

In *Chapter One* we have described environmental crises in Australia and noted the uniqueness, fragility and vulnerability of its environment. In *Chapter Two*, we see that the Australian public policy focus on development remains a primary cause of ecological degradation today. Walker argues that economic growth has driven the intensification of resource exploitation and that only palliative policies have been developed to address environmental impacts. In *Chapter Three*, Kate Crowley interrogates politics as the primary cause of climate policy failure by reviewing the thwarted Carbon Pollution Reduction Scheme proposed by a prime minister who was deposed for its poor handling. *Chapter Four* by Mark Diesendorf documents lost opportunities within energy policy and the constraining role that Australia's reliance upon coal as an energy source, and the maintenance and enhancement of its dominance, has played.

Following these chapters, the focus turns in *Chapter Five* to natural resource management, with Geoff Cockfield considering the ongoing struggle to develop policy solutions through the lenses of liberalism, politics and policy goals. Surely the sorriest case of environmental policy neglect is the Murray-Darling Basin, which is examined by Dan Connell in *Chapter Six* in terms of its century of decline and the failed policy efforts to stem disaster. In *Chapter Seven,* Geoff Wescott looks at Australia's failure to develop effective, integrated management plans for its 70,000 kilometres of coastline, and for its exclusive economic zone which is one of the largest in the world. Turning to national and peri-urban water policy, in *Chapter Eight* Michael Buxton examines institution-building and claims that have been made for the benefits, yet to be realised, of a market-based approach to water management.

In *Chapter Nine,* Michael Howes and Ayşin Dedekorkut-Howes consider the preparedness of coastal Queensland to adapt to the record floods and storms of recent years, which are predicted to become even more common with climate change. Policy learning, institutional capacity and preparedness for the projected increasing incidence of catastrophic bushfire are examined by Dave Mercer in *Chapter Ten* in the wake of recent disasters in the Australian Capital Territory and in Victoria.

In *Chapter Eleven,* Mark O'Connor puts over-population on the map as possibly one of Australia's most neglected environmental issues, where policy failure is endemic and where the myth of 'boundless plains' to share endures in practice. Then, of critical significance, given the litany of Australian environmental policy failures, is the right to dissent and the role that is played by environmental non-governmental organisations. Thus, in *Chapter Twelve,* Joan Staples considers the changing history of the relations between the government and the Australian environment movement and the need to strengthen pluralistic democratic advocacy practice.

The conclusion reviews the issues and themes raised by these chapters, reflecting on their common characteristics. It then suggests an overall framework, incorporating a political economy perspective, for evaluating policy failure.

Chapter 2

Australia's construction of environmental policy

KJ WALKER

Introduction

Humanity's dependence on the natural environment means that environmental variables must be placed first in understanding human socioeconomic systems. Economists in particular tend to assume that economic factors are fundamental. But no economic system can survive without inputs from nature: at the most basic level, energy and raw materials. Frequently, non-human organisms – bacteria, plants and animals – are needed to make the inputs available for human use.

Modern environmental problems hark right back to dilemmas voiced, but not resolved, by the political philosophers of the seventeenth to nineteenth centuries. Are there limits to the carrying capacity of human societies? What rights do humans have over nature? Are all humans equal? Can everyone expect a decent living? How can humans avoid damaging conflict?

Often, the answers to these questions are intertwined in ways less obvious three centuries ago. For example, the trade-off between population and quality of life: over-population leads not only to scarcity and cut-throat competition, but as it develops can cause congestion and other inconveniences. Human 'rights' over nature, it turns out, are not simply a pleasant conceit but a dangerous doctrine fraught with risk when natural systems are disrupted and human existence threatened. The possibility that there might not be enough to go round poses stark choices between a 'communism of famine' or an equally stark division between wealth and poverty. This latter, it turns out, is itself a threat to ecosystem stability and hence human survival.

Governments have tended to shelve these problems, evolving policy incrementally and pragmatically and ignoring poor 'fits' with reality. As recently as the last quarter of the twentieth century, 'responsible' politicians felt free to pooh-pooh suggestions that ecologically unsafe practices – from forest clearance to genetic modification to nuclear energy – could have adverse impacts on *human* societies. Widespread acceptance of the neoconservative and economic 'rationalist' world views adds further serious internal inconsistencies to an already confused 'philosophy' of human omnicompetence.

The dominant view of the natural environment, which underpins human economic activity in modern times, is due to the seventeenth century English philosopher John Locke (1632-1704), who regarded nature as useless unless exploited for human purposes (Macpherson 1962).

The Lockeian view has serious pitfalls. Locke's theory of property was based on the idea that property could be appropriated from nature; it did not take account of prior use by 'savages', nor, of course, of the natural biota. In England, Locke's homeland, where human occupation of the land had already transformed it irrevocably, this meant that resources were already scarce and had long since been allocated among the human population. Britain was a class society, with sharp differences in wealth and power. But at a time when European empires were expanding into 'uninhabited' regions, a justification had to be found for the theft of lands and resources. This was achieved by postulating that societies without European civilisation were in a 'state of nature' which preceded human society. Accounts of the origins of society, the ownership of property or its relationship to the natural environment depended on this conjecture. Resources in the 'state of nature' were assumed to be infinite and there for the taking (Macpherson 1962). In actual fact, of course, human societies evolved quite differently, and many of the 'vacant' lands were long since occupied and exploited, while available resources were often restricted (Diamond 1991, 1998, 2005).

This is most important from an ecological point of view because the assumption carries over into classical economics, treating natural resources as inexhaustible. But while humans often *behave* as if resources are infinite, they remain restricted in reality.

Australian laws on land settlement tracked the Lockeian view closely, mandating 'improvement': clearing vegetation, irrigation, substitution of 'productive' species, and attempting to 'make over' the land in the image of Britain. This both required and excused the abandonment of Aboriginal techniques of husbandry, which were more conservative of resources. Absentee landlords, and in recent times foreign multinational firms, have long dominated public policy.

Public policy in Australia since 1788 has been persistently developmentalist. This régime has been remarkably durable. It is grounded in nineteenth-century notions of 'progress', but applies it to an ecological system in which climate, soils, flora and fauna are all incompatible with it to a greater or lesser degree. Geared to European expectations, it ignored or belittled environmental constraints. The resulting pattern of development was subsidised by the running down of Australia's natural resources: minerals, soil, water, forests and native species. This highly institutionalised pattern inhibits environmental sensitivity and, consequently, sustainability, despite a 200-year history of criticism on many grounds (Birch 1975; Bolton 1992; Boyden, Dovers & Shirlow 1990; Lines 1991; Pyne 1992; Smith, D. 1990). Nowadays, it risks destroying the future.

Thus an environmental perspective on Australian public policy necessarily looks very different from more conventional accounts because it ignores much that is irrelevant while refocussing on issues previously considered trivial or peripheral.

Explanations of Australia's developmental pattern have tended to attribute failures to remoteness and aridity, and successes to a rich endowment of natural resources (for example, Blainey 1963). But these explanations are incomplete and *ad hoc*: they need a broader geopolitical and ecological context.

Statist developmentalism

From a very early stage, Australian governments subsidised economic activity to promote development, often in response to local political forces as well as imperial policy and domestic development ideology. Attempts to remove such benefits often provoked violent opposition from the pastoral, industrial and financial sectors (Cochrane 1989; Connell 1977). To avoid such conflict, undifferentiated economic growth was promoted through tax relief, services at low prices, free land and other inducements. Colonial governments borrowed from Britain, investing the proceeds in infrastructure, especially telegraph lines and railways, to stimulate growth. Their control of Crown land enabled them to stimulate agriculture and pastoral growth through generous lease terms and settlement schemes. They created marketing boards and engaged in overseas promotion of Australian products. From the late nineteenth century onwards, they set up government-owned enterprises to supply services such as post and telegraph and rail transport. Industrial and governmental interests became interlocked (though not necessarily identical).

This pattern of development is a form of *statism*, in which government takes a major, often dominant, role in development and economic management. It has been called 'colonial socialism', a term derived from the prejudices of 'free market' economics (Butlin, Barnard & Pincus 1982, p. 13). But that term was a misnomer, for socialism's traditional association with government enterprise is not its defining characteristic. The Australian development pattern was unequivocally developmentalist. It did not seek to further other socialist or welfare-statist objectives such as income equality, social welfare or a 'fair go' for working people. It was well established long before the rise of a strong labour movement, and decades before the Labor Party had any significant input to policy; to call it socialism is consequently inappropriate. Given the strong role of government, the term 'statist developmentalism' is more apt, capturing the Australian state's deep involvement in development (Walker 1994, pp. 249-50).

Statist developmentalism has gone through several stages in a changing economic environment. But the basic pattern is persistent, dictating a 'correct' route to development based on ingrained nineteenth century attitudes and ideas. Even present-day 'state-assisted marketisation' is sponsored and subsidised by government (Bell & Head 1994, p. 21; Davis 1996). As Cockfield (Chapter 8, this volume) observes, it has had very limited success in most public policy fields. Undifferentiated, 'cancerous' economic growth, careless of ecological impact, remains a tenet of faith, leading to avoidable conflict between 'growth' and 'environment' – both crudely defined.

Colonialism

Australia's colonial past is not unique. Europeans have invaded and dominated other societies for centuries. From the fifteenth century on, the Portuguese, the Spaniards and later the Dutch sought the spice islands of the East Indies. Politico-economic control of civilisations in Asia and the Americas followed (Columbus wasn't looking for Las Vegas: he was trying to reach China). Colonial possessions were exploited for their produce, especially spices, metals and, before the Industrial Revolution, textiles. Later, plantation agriculture was introduced in order to produce raw materials for European manufactures. These included indigo (for blue dyestuff), rubber, coffee, tea and cocoa, plus tropical foodstuffs such as bananas and pineapples.

But Europeans also settled 'underpopulated' regions of temperate climate, displacing the indigenous population and introducing the crops and animals typical of Europe and the Middle East. Grains such as wheat, barley and oats were accompanied by cattle, sheep, goats and pigs. This has been described as 'ecological imperialism' (Crosby 1986). Australia, virtually the last to be settled, was the least tractable in biophysical terms.

> *The development of colonial Australia was closely tied to the needs of the British Empire and took place within a capitalist world economy where the 'core' industrialised countries of Western Europe invested in 'peripheral' states whose role was to provide cheap raw materials and foodstuffs, a market for manufactured products and profitable avenues for investment of surplus capital (Frawley 1994).*

'Opening up' land for European settlement and modes of production such as agriculture, pastoralism, mining and forestry was accompanied by violent displacement of indigenous hunter-gatherer societies – both in the New World and in Australasia (Lines 1991). In Australia, whaling, sealing and the exploitation of exotic timbers such as cedar and sandalwood were followed by wool, minerals and wheat. Imperial trade incorporated the new societies as suppliers of food, fibres and minerals.

Western Europe has reliable annual rainfall, deep alluvial soils and a temperate climate. Environmental exploitation there relied on predictable returns and a consequent expectation of steady annual incomes. Symptoms of ecological abuse were less noticeable and more easily repaired. In Australia, settlers encountered a radically different climate in which drought was the norm, with flora and fauna they did not understand and failed to use sensitively or sustainably (Smith, D.

1990; Boyden, Dove & Shirley 1990; Bolton 1992). They attempted instead to 'Europeanise' it.

Colonial but hardly socialist

The Australian economy grew rapidly during the 'long boom' from the 1860s to 1890. The gold rushes of the 1850s began a 'rush that never ended', as more minerals were discovered and exploited. Forests were felled for fuel, and landscapes increasingly scarred. 'Closer settlement', irrigation and land clearance emphasised physical development, and were insensitive to Australia's fragile and unique natural environment (Powell 1976, Bolton 1992).

Major changes in flora and fauna transformed the landscape. Clearing of land, in particular native scrubs such as brigalow, continued up to the 1990s; it played a large part in soil loss (Lines 1991; Ratcliffe 1948). The impact on native fauna was devastating, and worsened by deliberate introductions (Marshall 1966; Rolls 1969). Cities devoured, and continue to devour, scarce fertile land, while polluting coastal waters, air and littoral heavily (Smith, GJ 1990; Beder 1989; Whitelock 1971).

Government service provision took place both through departments under political control and statutory corporations – government enterprises administered by nominally independent boards – from the 1880s (Mann 1981; Hancock 1930). State assistance to rural industry, from infrastructure provision to marketing schemes and purchase of ailing mines, subsidised and worsened these impacts. Drought 'relief' helps perpetuate the myth that bad years are exceptional; continuation of unsustainable farming practices then increases stresses on soils and flora. Grandiose irrigation schemes – the Burdekin, the Snowy, the Ord – were undertaken without proper evaluation of their environmental impact, which was often severe. Many were white elephants. Land was opened up for inappropriate, water-hungry crops such as cotton and rice, storing up conflict for the future. These schemes then subsidised indiscriminate economic growth.

Dissent was not tolerated. While many critics were dismissed or ignored, serious and well-grounded criticism attracted outright persecution. The distinguished geographer Griffith Taylor, whose book *Australia* (1941, revised 1943) argued that much of the country was unsuitable for settlement, was hounded from his academic post in Melbourne and finished his career in North America. His views have since been endorsed by Davidson (1965, 1969) and Carden (1999).

Many of these activities, especially land clearing, resulted in ecological damage which in turn required – or will require – expensive remediation. There is an ongoing conflict between the Queensland sugar industry and the management needs of the Great Barrier Reef, primarily as a consequence of fertiliser runoff from cane farms. Water from the Snowy River scheme had to be diverted back to the Snowy River to restore its health. The over-allocation of water in the Murray-Darling system is at the heart of its ongoing problems. In some cases, such as the over-cutting of old growth forests – economically irrational in any case – the damage is irreparable within any realistic timescale.

Nor did extensive government infrastructure provision automatically stimulate economic growth, let alone increase net welfare. Taxpayers frequently saw no significant return on their investment, and sometimes they were actually 'taken for a ride' by private investors. The dogged pursuit of 'development' was a main cause of disappointment. State infrastructure investment in advance of private interest or proved demand is essentially speculative. It courted all-too-frequent failure. Its precariousness, historically, was increased both by poor evaluation of projects before their acceptance and poor or non-existent assessment once completed. States were burdened with unremunerative railways, useless dams and politically untouchable subsidies to favoured groups – especially farmers.

The involvement of multiple governments and multiple interests in the community deepens the complexity. The more parties that are involved, the more likely it is that policies will be compromised and that the process of policy-making will be riddled with short cuts, *ad hoc* approaches and the exclusion of parties with legitimate interests.

The march of economic irrationalism

Australian public policy underwent an apparent change from the 1980s. Any attempt to insulate the economy from global trends was abandoned in favour of openness to global markets. New Right rhetoric, repeated unanimously, uncritically and *ad nauseam* by the mass media, underpinned the change (Pusey 2010). Part of the trend to globalisation of the world economy, this saw the mass movement of wage labouring employment to low-wage countries, and a vast expansion of contract, sweatshop and casual labour, heavily curtailing the political power of the traditional labour movements. Giving priority to economic criteria – and especially to corporate profit – seriously undermined previously promising environmental policy trends.

These trends were widely seen as marking a revolution in public policy, and in particular as an absolute contraction in the role of government. And yet, although the centre of gravity of Australian politics moved significantly to the right, abandoning much that had once made it humane, government remained strong.

In the 1980s and 1990s, state enterprises ranging from banks to communications were sold off, inspired by a drive to withdraw government from directly marketed services. The motivation was allegedly to stimulate competition, uncritically accepted as bringing reduced costs to consumers. However, the underlying 'New Right' economic philosophy, which had become especially dominant in the Commonwealth Public Service, paid little attention to verifying this claim and none to environmental costs or sustainability. Nor did the extent of government commitment to capital investment greatly change, despite the rhetoric of 'smaller government'. Communications and road building expenditure, for example, grew, although social services were generally cut. Government remained willing to contemplate huge infrastructure expenditures, especially in 'sunrise' technologies such as digital communications.

This ambiguous trend also saw the partial withdrawal of government from regulation, especially in areas such as banking and finance (Bell & Head 1994). Though the segmental nature of the economy selectively constrains the capacity of government to direct and control, it points at the same time to the continuing importance of government, both in setting overall goals and objectives (although within clearly-defined constraints arising from global forces) and in creating the framework for the 'new' economic order.

But dramatic changes in economic policy may have very little impact on development policy, at most altering its emphasis. There remains an element of pump-priming - now called 'neo-Keynesian' - and continuing acceptance of the 'need' for government to provide infrastructure. Focussing attention on the decline of old infrastructure and subsidy obscures the rise of new. Road transport, for example, continues to be massively subsidised; its social costs are ignored, and the application of 'level playing field' criteria to transport is virtually never discussed. Attempts to improve the balance of payments by exporting larger volumes of the traditional unprocessed exports have led to massively increased environmental impacts, especially when increased volume attempts to make up for falling prices. Alternative strategies involving greater intrinsic value and higher employment - such as production of high-value cabinet timbers rather than woodchips from native forests – have not been adopted; 'value added' tends to mean large, foreign-owned plants such as pulp mills. The new policy direction continues to treat natural resources as free goods and ecological damage as trivial and reversible. This points not to the supplanting of 'statist developmentalism' by market-oriented philosophies, but to its persistence in a changed form.

The Hawke-Keating Government of 1983-1996 abandoned Labor's commitment to social justice, and embraced much of the 'economic rationalist' or neoconservative policy agenda. 'Wage flexibility' - that is, lower wages for workers but not for bosses - was espoused, against a never-redeemed 'accord' promising better conditions once economic conditions improved. Some progress was made on the environment - the blocking of the Franklin-Gordon dam project in Tasmania for example - but Keating in particular was dismissive of environmental concerns.

Most significantly, worried by uncertainties over oil supplies and fear of balance of payments problems, the Hawke-Keating Government decided to base both energy policy and balance of payments on the exploitation of Australia's abundant coal reserves. Coal would be dominant in electricity generation and would become the most important mineral export. Research would be directed to ways of mitigating the greenhouse impacts of coal; as Diesendorf (Chapter 4, this volume) remarks, this led to a reduced emphasis on research into renewables. This policy stored up trouble in at least three ways. First, massive exploitation of coal would increase greenhouse gas emissions, both globally and in Australia. Secondly, as Bartlett (2006) has demonstrated, the enhanced rate of extraction leads to exponential depletion, reducing the expected life of the resource from hundreds of years to mere decades. And, thirdly, the strategy of reliance on coal led to economic distortion, the rejection of alternative and more viable policies, and the flight of promising technologies overseas.

One consequence was an increasing dominance by miners over public policy. Their demands for access to mineralised land, without consideration of other values such as ecological integrity, spiritual values, landscape or amenity, had become common from the 1970s. They posed a serious threat to Aboriginal lands in particular. Additional demands, designed to tailor immigration and population policies to the needs of mining companies, included the free import of skilled labour, a large pool of contract workers – all without job security - and demands for a 'bigger Australia' via continued mass immigration in addition to an accelerated natural increase. There was no commitment to education or training; previous 'reform' of the education system had left a serious shortage of skilled labour even without the demands of mining. And, the issues of safety, rehabilitation and pollution were simply not addressed.

The Hawke-Keating Government's commitment to coal was perpetuated by the successor Howard Government. Import of foreign labour was facilitated, research into alternative energy sources was truncated, and massive investment in carbon sequestration (CCS) was espoused. CCS envisaged capturing carbon emissions from coal-fired industries – predominantly power stations – and storing it underground in 'safe' rock formations. The technology was unlikely to be available before 2020; economic and energy costs were undetermined, and the availability and accessibility of suitable rock formations was unknown. (A similar problem had arisen in the 1970s with plans to store nuclear waste underground; it was never resolved.) CCS was important solely because it offered the coal industry the prospect of continuing demand – albeit through a technology which was undeveloped and untried, and which contained as yet unknown pitfalls (Diesendorf 2007, 2009).

Labor won the 2007 federal election with a clear majority in both Houses of Parliament. For only the second time in Australian history, an incumbent prime minister lost his seat in the House of Representatives, as the former PM, Howard, was ejected from the formerly safe seat of Bennelong. Rudd, the new PM, had campaigned on climate change, and there was a widespread expectation of decisive action. Instead, there were long delays in formulating policy. What finally emerged was a carbon trading scheme (ETS) which 'compensated' the worst polluters most generously and perpetuated the commitment to CCS. This proposal was rejected as useless in the Senate. Although Rudd had gained power on the promise of firm action on greenhouse issues, there was little visible change from the Howard policies. The weak cap-and-trade (ETS) scheme was followed by 'leadership from behind' at the Copenhagen conference in 2009. Then, late in 2009, Rudd was suddenly overthrown by his deputy, Julia Gillard, and Labor went to the polls shortly afterward. They lost control of the Senate, and only retained government through the support of three Independents and a Green in the Lower House. By early 2011, there was still no sign of a comprehensive, decisive policy on global warming, although a carbon tax had been announced that was to commence in 2012. The presence of active doubters of the significance of global warming within Labor, and the evident fact that Green pressure was driving the policy agenda, led to speculation that the party system was changing: two conservative 'dinosaur' parties were coming increasingly to confront a Green trend, the dynamic of which was at an angle to the conventional right-left division, and which had a novel agenda of its own.

Quarry Australia

Guy Pearse, a former insider in the Howard Government, caused a furore in 2007 with his book *High and Dry: John Howard, Climate Change, and the Selling of Australia's Future* (Pearse 2007). It was a passionate view of resource and climate change policy under the Howard régime. Its scathing exposé of the self-described 'greenhouse mafia', an organised group of representatives of major mining and energy-consuming industries, underlined the dominance of big energy and mining corporations – not only in policy but also in public debate. It offered a compelling account of the mining lobby's use of influence, pressure politics, and their direct access to Howard's inner circle. His essay 'Quarry Vision' (Pearse 2009) extended this condemnation to the Rudd Government's failures. He drew attention to derisory controls on foreign investment, and the ignoring of adverse economic and social effects of mining. While the revelation that the conservative side of Australian politics was deeply entwined with big capital was hardly novel, it underlined the reasons why environmental policy had gone so sadly adrift.

Pearse attributes the dominance of the 'greenhouse mafia' to reliance on mineral exports as a major support for the balance of payments. This had the effect of subsidising domestic consumption from the proceeds of exporting minerals, typically as unprocessed raw materials. It continued the shift in emphasis from agriculture and pastoralism to minerals. Its environmental significance lies in its substitution of *depletable* resources – minerals – for renewable commodities. Politically, it gave the miners enhanced political leverage. As long as minerals had remained a fraction of total production (and especially exports), neither depletion rates nor their impact on the economy were significant (Blainey 1963; Birrell, Hill & Stanley 1982). With growing international interest and investment in Australian mining, overall output expanded and coal mining for export grew explosively. Mining was attractive to resource-rich states such as Queensland and Western Australia, promising expanded revenues for minimum effort. Coal in particular was popular because Australia's vast reserves were thought to be virtually inexhaustible.

This fitted in well with neoliberal thinking, which emphasised unplanned economic growth and ignored long-term consequences, let alone ecological impact. Abundant coal reserves permitted rapid and expanding exploitation without concern for depletion, while the proceeds, including the meagre job creation involved, could be represented in a positive light. The fly in the ointment, of course, was that increased exploitation of *any* finite resource must increase the rate of depletion. Bartlett (2006) demonstrated that depletion of seemingly abundant resources can be startlingly rapid, even under quite conservative assumptions; in Australia's case, coal depletion times would reduce from the often-quoted centuries to mere decades.

Global warming and the associated policy measures seriously challenge high-volume mining and export. Abroad, exports are threatened by reductions in demand for coal in particular, but also by reduced demand for minerals in general. At home, increased attention to pollution, land degradation, accident risk, and so on, all threaten the coal industry and coal-fired electricity. The economic 'boom' is bought at the cost of market-induced instability and massive demands for – necessarily temporary – investment in infrastructure and training. Inflation, labour market imbalances and stresses on education then follow.

For a mineral-supported economy in which coal mining and exports are essential props – like a 'security blanket' – there are two paths out of this problem. One is policy change, to accommodate a more realistic worldview; the other is denial and hostility to global warming science.

The closed policy environment of the Howard Government saw the fixed, dogmatic pursuit of neoconservative ideas, the narrowly circumscribed choice of policy advice, and deceitful and manipulative relations with press and public (Hamilton & Maddison 2007; Pearse 2007, 2009; Coghill 1987). Nor were other actors innocent of deceit and bias. The press and other media acted more as a pressure group than an 'honest broker', persistently giving prominence to neoconservative views to the exclusion of others; rational discussion was choked off or forced back into less influential channels (Beder 1997). The effect was to bias public opinion destructively, by denying it relevant information, serious analysis and alternative points of view.

Capture? Or complicity?

Pearse and others attribute the Howard Government's behaviour to capture by major industries, especially mining. Capture in turn is attributed to the politicisation of the public service. But, firstly, there is an interesting paradox here: governmental grip on policy has not been significantly relaxed, so statist developmentalism has persisted even though it now serves sectional interests. This was not a case of 'capture', which requires the imposition of policy by powerful non-governmental agents. Government cooperated willingly and as a matter of policy. And, secondly, the intertwining of government and the corporate sphere is far more extensive, and far more complex, than a mere dominance by miners, however important individual governments may have regarded mining.

Challenges to statist developmentalism

Statist developmentalism's *distributive* approach to public policy is geared to big 'projects'. In the absence of private capital, governments became sponsors, often making substantial investments. Traditional beneficiaries were contractors engaged in construction work and their employees, or recipients of subsidy, e.g. farmers benefiting from cheap water or fertiliser. Nowadays, large multinational corporations have also benefitted. Frequently, for political reasons, state governments in particular become committed to grandiose schemes for which they have few resources. Often, like the Burdekin scheme in Queensland, these would be undertaken in dribs and drabs, with only a little capital being allotted in each year's budget. As a consequence, states became committed to paying interest on loans for schemes which would not yield revenue until far in the future, if at all. That, in turn, increased the tax burden and tied up funds that might have been devoted to other more useful purposes.

Another consequence has been large-scale corruption, which has been remarkably common – especially in New South Wales and Queensland. Because government controlled the granting of contracts and, frequently, the appointment to jobs as well, it was easy to trade favours.

Distributive policy patterns begin to become a liability when resources are no longer abundant and cheap, and consequently there is competition, not only for the resources themselves but also from alternative uses ranging from other 'cleaner' industries to recreation and wilderness. When the environment can no longer be freely exploited and the end of the formerly inexhaustible resources is in sight, investment no longer expands employment. Instead, the biggest and most capital-intensive projects create few jobs and the most disruption. As the effects of development increasingly impose diseconomies on the population, the benefits are no longer 'self-evident', and the associated policies fall dramatically in popularity. In short, when 'ecological scarcity' begins to bite, and the resource régime has to change from distributive to regulatory, statist developmentalist policies can become increasingly irrelevant and even harmful.

Policy obsolescence poses policy dilemmas for government, especially in regards to the need for novel approaches to economic and environmental policy. Serious attempts at a transition are lacking; the 'Greenprint' for Tasmania, the sole comprehensive exploration of the problem, was ignored and shelved by the very government that had commissioned it (Lowe 1991). Distribution, not regulation, remains the dominant theme in modern Australian politics.

Statist developmentalism is as much a state of mind as a development strategy. It embodies a series of assumptions, among which the most important are: (1) that development is imperative, (2) that development is popular, and (3) that development has self-evident advantages. There are still advocates of large-scale 'development' through massive government investment in arid country areas. In the face of global warming, this has extended to advocacy of northern development in Australia, perhaps in ignorance of the barriers. This mind-set ignores evidence that development damages ecologies and diminishes amenity for the population at large. Instead, it assumes that ecologically rational policies will be costly and will eliminate jobs, setting up an unnecessary and damaging conflict. The outcome is frequently that policy-makers are unwilling to consider perfectly viable options. Coupled with deeply entrenched prejudices and interests in government departments, such attitudes explain the long persistence of ecologically, economically and even politically irrational policies.

There is already substantial agreement on many of the policy measures necessary to deal with the threat of global warming. There is less sense of urgency and less interest in stabilising Australia's ecology, though its problems are as pressing and are intimately linked. Both can be characterised as 'ecological scarcity'; that is, not simply scarcity of resources, but also scarcity of options, as the 'funnel' narrows, diminishing the viable choices.

Major policy measures already suggested include the withdrawal of subsidies to major polluters - aluminium, coal mining and fossil fuel power generation - and entrenched rural industries with major detrimental impacts: cotton, rice, and extensive grazing in unsuitable regions.

Recent history suggests that such measures would provoke major disinformation campaigns and vehement opposition, especially from conservatives and major newspaper chains. The Murray-Darling Commission's minimal proposals to restore flows in the Murray-Darling Basin (MDB) provoked violent and ill-informed reactions from farm interests, inflamed by an irresponsible Opposition. Announcement of a carbon tax generated a barrage of misinformation and rejection, again inflamed by the Opposition; this is particularly ironic since the same measure had already been proposed by them and would in any case be unavoidable were they to regain government. The amounts spent on disinformation campaigns also suggest that the culprits can well afford both to do without subsidies and to pay more realistic taxes. It appears, for example, that the mining industry, which is largely foreign-owned, set aside $100 million to campaign against the proposed super-profits tax on mining - of which $7 million was spent in 2010 (Lee 2010; Schofield 2010). And this does not take into account money paid to propagandists and 'think'-tanks (Carey 1987; Beder 1997).

Major battles loom as policy changes for greater sustainability eventuate. There are numerous potential flashpoints. Energy suppliers, such as oil multinationals and coal miners, are likely to seek to retain control over their markets. Some (BP, for example) have a history of attempted repositioning to exploit alternative (but monopolisable) technologies. Others have tended to resist change; a major development in northern NSW and Queensland has been the attempt to develop 'coal seam gas' as an alternative energy source. This involves fracturing coal seams *in situ* by explosives, and then capturing the volatile gases released. Claims that toxic chemicals have appeared in local water supplies, and massive resistance by local farmers to disturbance of their land, have generated formidable opposition. The technology is by no means proven, nor are costs and likely duration known reliably. Coal seam gas appears to be another example, like CCS, of the dogged pursuit of fossil energy despite the growing viability of renewables.

Another flashpoint arises after 'peak oil', and where electrical energy needs are rapidly increasing. This situation has possibly occurred already, and certainly will in the near future. Where dependence on fossil fuel is high, prices could rise very sharply, creating 'windfall' profits for suppliers. Electricity suppliers could be especially resistant to decentralised power supply options, such as solar photovoltaic (PV) or wind, which could threaten supplier control of energy. Both predicaments could generate strong pressure for delaying or rejecting even palliative policy measures.

Such conflicts are tests for statist developmentalism. A policy régime wedded to powerful vested interests might well be too inflexible for significant change. However, if a determined government were to pursue effective policies - and, importantly, maintain popular support - then statist developmentalism might undergo significant change, retaining both the statist element and a developmentalism which necessarily differed significantly from what had gone before. The other possibility - less probable, perhaps - might be its replacement by a different régime. If continuing pursuit of New Right fantasies is excluded, one possibility might be a more responsible policy stance, sensitive to its environmental impact, dedicated to the welfare of the community, and capable of using the available talent creatively.

For the immediate present, the major policy deficiency globally has been the lack of regulation, not merely in environmental matters but also in such fields as banking. With greater governmental participation in the economy than in some other 'western' democracies, Australia might be in a better position to extend and fine-tune regulation. However, the litmus test would be a determination to change policy comprehensively.

The question of growth

A major component of the environmental debate has been the question of material *limits to growth*: that is, whether the materials and energy needed to support human technology will become so scarce that civilisation itself will collapse. Obviously, since the earth is a small and finite planet, if human populations grow sufficiently large it will certainly be impossible to feed, clothe or house them all. While technology can raise carrying capacity, it remains dependent on inputs from the natural environment. These can be reduced by efficiencies, but they cannot be eliminated with any known or conjectured technology. Very rapid population growth could bring on crisis very abruptly: a 'tipping point'. Most arguments against this view are not designed to deny the possibility of ultimate limits, but to show that they do not pose a serious threat in the foreseeable future.

A common claim is that resources will never become scarce because they can be substituted; that is, as one thing grows scarce, another will replace it. This line of thought assumes that substitutes will always be available, and, so long as some sort of substitute is available, things never become absolutely short. But substitutes for some materials are very hard to find; and exponential growth in demand typically leads to scarcity and high prices rather than simply running out (Walker 1979). The greater the demand for a scarce material, the greater the likelihood that growing scarcity will precipitate supply crises and hence economic turmoil. Diamond (2005) has underlined the importance of such crises for the failure or survival of societies in historical times. Typically, the rulers fail to respond creatively to decreasing productivity, and a collapse, marked by material scarcity, social conflict and economic paralysis, eventuates. There is no reason to suppose that modern societies would be immune: a global economy exposes all participants to failure in a single sector, as the 'global financial crisis' of 2008 demonstrated. Australia's dependence on high technology agriculture, which uses imported fuel and is dependent on foreign markets, makes it particularly vulnerable.

Policy adequacy

Government policy is 'adequate' when it can be shown to have identified a problem in a timely fashion, and to be tackling it at a rate sufficient to stem it, reverse it, or at least manage it. Unsuccessful policy is by definition inadequate. The problem raises numerous difficulties, from underlying philosophies to criteria for measurement (Walker 1995).

Political systems are always in states of dynamic equilibrium. Even when 'stable', they are in continuous flux. The critical issues include both the direction and the rate of change. Adequate policy must therefore address both fundamental and long-term problems, as well as any programs designed to engage them (Lowe 1994).

But environmental policy must take account of four important characteristics of ecological systems:

1. *Invisibility*. Many problems grounded in ecology are *invisible*. This can lead to ignorance. For example, some pre-scientific societies were unaware of the consequences of their policies: the kingdoms of ancient Mesopotamia irrigated extensively, but their soil fertility was destroyed by salination. This history has been replicated in the Punjab and in Australia. In modern times, the use of chlorinated fluorocarbons (CFCs) was banned only *after* the discovery of atmospheric ozone holes. This was a failure of knowledge coupled with a failure to regulate. With thousands of new compounds - and nowadays genetically modified organisms - being introduced to the biosphere annually, similar problems are sure to recur. Rapid technological change, with exponentially multiplying problems, poses a particularly serious threat.

2. *Irreversibility*. There is serious concern about 'tipping points': junctures at which critical ecological chains or cycles - e.g. carbon, oxygen, nitrogen - overload and break down, leading to runaway change. This fear is especially strong in connection with greenhouse gases and global warming. Species loss, affecting not large mammals but small and unconsidered species, is another risk. As each tipping point is passed, the options facing policy-makers diminish, as an 'options funnel' of diminishing width increasingly constrains policy choice. In energy, for example, a transition to genuinely renewable sources may be progressively limited as cheap non-renewables run out - thus, increasingly we may be faced with 'you can't get there from here'.

3. *Indivisibility*. Following the publication of Hardin's notorious essay, the idea that only benevolent dictatorship could solve the 'environmental crisis' briefly became widespread (Hardin 1972). It was heavily criticised and is rarely advocated seriously nowadays, mainly because the

invisible claw of monopoly markets – popularised as 'neo-conservatism' and 'economic rationality' – has served the needs of conservative elites better. But the short-termism of neo-conservatism, and the pervasive insecurity of the boom and bust cycle it creates, disrupts the lives of ordinary people and concentrates wealth and power in the least responsible of hands. The underlying issues of equality, equity, fairness and the management of common goods have remained unresolved in conservative thought since the time of Locke.

Most environmental goods are *common goods*, as they have at least an element of jointness in their provision. That is, if provided for one, they must be provided for all (Laver 1981, pp. 39-71; Ostrom 1990, pp. 1-23; Walker 1994). Liberal/conservative political thought is ill-equipped to cope, because it is geared to justifying exclusive rights to property.

4. *Urgency*. In the face of rapid change, 'make haste slowly' is suicidal. This was very clearly illustrated when the Rudd-Gillard Governments attempted, first, to create an ETS in 2007-2008, and then to impose a carbon tax in 2011. Both carbon taxes and an ETS had already been adopted in Europe – 30 nations had a carbon tax from 2005, and three more had legislated for one by 2011 – but the Howard Government's delay had significantly increased the difficulty by 2007. Vested interests, especially in the coal industry, had demanded the incorporation of massive compensation in the ETS, the effect of which would be to underwrite continuing carbon emissions. The ETS legislation was defeated in Parliament, crucially, on the ground of its ineffectiveness (Crowley, Chapter 3 this volume). Both a carbon tax and an emissions trading scheme, as well as 'direct action' by banning new polluters and penalising existing ones, are necessary, but delay and prevarication steadily increase the cost.

Especially when multiple problems must be tackled simultaneously, emphasis must fall on holistic solutions which integrate several perspectives.

Transforming policy

Policy prescriptions for coping with environmental change date back at least to the 1950s (Brown 1954). There are many more recent contributions (for example, Lowe 2009; Diesendorf 2007, 2009). Here, a few central considerations are explored.

Numerous pitfalls face policy transformation, and some shibboleths must be discarded. For example, many politicians talk about 'balance' in policy. What they mean, in most cases, is a 'balance' between competing demands. But situations can and do arise in which no 'balance' is possible: there is an imperative – frequently ecological – that limits the space for compromise. A conspicuous example is global warming: if widespread ecosystem collapse is to be avoided, then atmospheric CO_2 concentrations must be limited. There is already widespread agreement among scientists on the amount of limitation required, but policy proposals - both national and international - fall significantly short of desirable cuts to CO_2 levels, to name only one variable. Such policies are inadequate: any 'balance' they offer is a recipe for disaster.

Another pitfall is the practice of 'offsetting' polluting activities by supporting other activities - often in remote locations or overseas - which are thought to make up for polluting emissions by the parent activity. The most common of these is tree planting, which is thought to exploit the ability of trees and other greenery to 'blot up' CO_2. If the 'blotting up' is remote from the pollution source, of course, its immediate utility may already be in doubt. But a more serious problem is that recent research suggests that the ability of newly planted trees in particular to absorb CO_2 may be more limited than was thought (Beale 2010; Norby *et al.* 2010). In these circumstances, offsets are not even palliatives.

When *urgency* is a consideration, the long term, especially, becomes paramount. In 2010 and 2011, irrigation farmers in the Murray-Darling Basin expressed outrage at proposed buy-backs and reductions in water allocations. But they had known for years that the system was in trouble, and a primary cause was over-allocation of water. This is an example of a multi-person 'Prisoner's Dilemma' game, in which one or more players is attempting to corner all the gains (Laver 1990, pp. 21-71; Walker 1994; Ostrom 1990). The collapsing societies described by Diamond succumbed very often to the same virus. *Cooperative* solutions embracing all those affected are essential to solving such problems.

In such cases, it is of critical importance to identify the necessary outcomes, and to discard policies which will yield half-measures or failure. In short, transformative, and not palliative, measures should be sought.

If, for example, grower outrage in the MDB were to drive a policy of token reductions in water allocations, leading to 'environmental' flows below the necessary minimum, that policy would be palliative and would likely fail, with greater damage to the Basin - and the objectors themselves - than a more equitable and radical alternative.

Palliative measures are often short term. Replacement of coal by gas for electric power generation is currently canvassed as a way to manage a transition from coal-burning to a future fully renewable scenario. But substituting gas for coal is not a durable solution since gas supplies will also run out in due course, and gas is also a source of greenhouse emissions. It is a short-term option as, even more than oil, it will suffer rapid depletion in the face of heavy demand. Even if adverse environmental impacts are ignored, it is a wasteful policy. The main reason for urging it is that, while dispersed technologies such as small- and medium-scale wind power and solar photovoltaic are immediately feasible, few successful *centralised* facilities - e.g. power stations - yet exist. Extensive decentralisation of electricity generation threatens 'business as usual' for power utilities geared to central generation and distribution. As Diesendorf (2007) and Jacobson and Delucchi (2009) show, renewable technologies exist and are entirely practicable, but they require radical changes to the existing system, as they are thoroughly transformative.

Outright dead-end technologies are those which create more entropy than they destroy, or are simply too expensive or too cumbersome to continue with. The prime example, of course, is the generation of electricity using nuclear power. Early predictions of power 'too cheap to meter' are now obviously laughable. Nuclear power stations have proved expensive to build, are prone to 'minor' – and rather too many 'major' – accidents, and pose unsolved and un-costed problems for decommissioning and waste storage. It is now widely accepted that, over their full life cycle from construction to decommissioning, nuclear power stations require more energy than they generate (Lowe 2011, Diesendorf 2005a, 2005b). In the face of other feasible sources such as wind, solar or geothermal, burning coal to make electricity is another dead end (Diesendorf 2007, Jacobson and Delucchi 2009). And the use of oil-dependent synthetic fertilisers to maintain soil fertility, when more conservation-oriented cultivation techniques are more suitable, is self-extinguishing.

These examples suggest overall policy rules. Dead-end policy regimes attempt to *dominate* nature, rather than working with it. The ill-effects are especially visible in Australia, where deliberate introductions and ecologically perverse practices have not merely destroyed native ecosystems, but have made it impossible to restore them. As urgency increases, and time to make changes runs out, policy benefits must be increasingly durable, based in long-term, transformative policy regimes.

There is also a 'sleeper' issue underlying not just durability, but the very stability of human societies. Marvin Harris, an anthropologist who played a major part in the development of the modern theory of the development of the state, has also underlined the importance of social equality. His model, *inter alia*, stresses the accumulation of social power and the development of hierarchy as important precursors of modern, stratified societies with sharp class distinctions. Concentration of power often went hand-in-hand with breakdown of ecologically stable communities. Populations in a state of ecological equilibrium would be forced by a controlling power to exploit their resources more intensively, at the cost of sustainability. Often, too, there was attendant pressure to increase population (Harris 1979; Diamond 1998). These forces are still in play today. The problem that remains is whether, with modern science and technology, societies which can maintain ecological stability and substantial equality can be devised.

Wilkinson and Pickett (2010) have recently drawn attention to well-documented links between social inequality and a wide range of disorders – some collective and others individual. Their conclusion is that 'equality is better for everyone', and that societies should aim for it. Significantly, they suggest that ecological health also depends on a substantial measure of equality: rich people waste resources, poor have no alternative but to destroy them.

Lowe comments that:

> *A stable society has to be an equitable one, because inequality – real or imagined – has always been a source of tension and instability. Widening divisions between rich and poor fuel resentment and social tensions. We need*

> *to change the national policies that are producing a less equal society (Lowe 2009, pp. 91).*

The decades from the 1960s on have seen a determined attempt by the political Right to dismantle institutions such as the welfare state, to downplay fairness and equality in favour of indiscriminate economic 'growth', and to impose their views on the whole world. Australia - once a world leader in social innovation - has been a particular victim of this trend, which is profoundly unpopular (Pusey 2003, 2010). As awareness of major environmental damage has increased, however, so too has an appreciation of the damage inequality and unrestricted growth can do. The evidence now seems clear: that social justice is an essential ingredient of ecological health.

Paradoxically, then, if the only viable survival strategy is one based on greater equality and ecological sensitivity, the survival of strong government institutions might turn out to be an unanticipated blessing. Statism without developmentalism could itself be transformative.

Chapter 3

Climate policy failure: Was Australia's CPRS more politics than policy?

KATE CROWLEY

The 'whole process of policy-making over the CPRS has been one of the worst examples of policy-making we have seen on major issues in Australia. I suppose it was never going to be easy, but the way it has broken down was extraordinary' (Garnaut 2009).

Introduction

One of the greatest environmental policy failures of the Howard Coalition Government (1996 - 2007) was its failure to tackle climate change, with the Prime Minister shackled to self-proclaimed climate change scepticism for over a decade. Climate change, or global warming, had been an issue of the highest concern to the Australian public for the entire time that the Howard Government was in office, and yet no effective policy was designed to reduce greenhouse gas emissions. The climate change policy failure of the Howard Government has been broadly criticised for failing to rein in soaring energy emissions, but also for stymieing the potential of alternative energy, and for protecting the interests of energy intensive industries. The Howard Government failed to act on the international or domestic stage for over a decade, so the challenge of reducing Australia's emissions steepened considerably by 2007. Opposition Labor leader Kevin Rudd appeared to appreciate the urgent need to act, prompted by the Stern report and severe drought, promising a carbon target, a renewable energy target and fund, a review by Australian National University economist Ross Garnaut into the feasibility of an emissions trading scheme (ETS), and the ratification of the Kyoto Protocol.

This sharp difference in approach on climate change was noted by a public that had been concerned by the Howard Government's inaction. In the lead-up to the 2007 election, Kevin Rudd showed considerable leadership. To investigate the implications for Australia of the Stern report, he held a national summit of the Labor party on climate solutions, and commissioned analysis of the efficacy of an ETS. Labor campaigned on an aggressive, innovative climate change platform, which forced significant concessions from the Coalition Government at the time. Both Labor and the Government committed to a national ETS (Labor's Carbon Pollution Reduction Scheme), but Labor committed to a target of 60 per cent emissions reductions from 1990 levels by 2050, and promised $3 billion in climate change funding over four years, trebling the Government's commitment. Labor's announcement of a twenty per cent renewable energy renewable target by 2020 drew a fifteen per cent target from the Government, which had previously proposed to abandon its earlier two per cent scheme altogether, and Labor under Rudd rejected entirely the Government's faith in nuclear energy (Crowley 2010).

By the 2007 federal election, it was clear that climate change was no longer simply an issue of the utmost environmental significance in Australia, but an issue of the utmost and critical political significance. After a decade in opposition, the Labor Party won Australia's first climate change election (Rootes 2008) and Rudd ratified the Kyoto Protocol as his first prime ministerial act. Climate change then drew unprecedented political blood for a national environmental issue in Australia. Over the next three years, climate change was to claim two opposition leaders and eventually the Prime Minister himself, who was deposed by colleagues as his first term drew to a close (Gettler 2010). The Labor Government would be punished for its bungled attempts under Rudd to establish an ETS, and to devise and implement workable climate policies that would make a noticeable difference. At the 2010 election, the Government then lost its majority and was forced to form a Government relying on independents and the Australian Greens, with carbon pricing a foremost concern. What follows is a critical review of this climate policy failure, here interpreted simply in terms of the Rudd Government's ETS's policy promise and its messy political demise, with an identification of constraining and facilitative factors involved in the climate action challenge.

The climate problem in context

Whilst the problem of climate change has garnered considerable attention as a scientific, moral and political challenge, there has been less attention paid – until recently – to the mechanics of successful or failed climate change policies or mitigation efforts. In Australia, there is a critical need to analyse the country's longstanding climate policy failure, perhaps best represented by the Howard Government. Is it Australia's carbon intensity that has so constrained successive Governments from acting and, if so, how could action be expected from other carbon-intensive nations such as Canada, China, Russia or indeed countries in the developing world? Or is it, as Garnaut (2010) now claims, simply a failure of leadership in Australia as leaders are unable to stand up to vested interests and restructure Australia's resource-based economy around a market-based price on carbon? Even though Australia's share of global emissions is less than two per cent, and its emissions reduction achievements are never likely to be globally significant, there are lessons to be learned from the politics of its failed policy efforts, and the failed ETS scheme in particular. These lessons regarding obstacles to domestic action to reduce emissions in turn may be scaled up and generalised, and may prove useful in a comparative context.

There are various imperatives for action on climate change in Australia that counter its insignificant share of global emissions and reinforce the case for an urgent pricing of carbon and for a low-carbon economic restructuring. For example, as Pittock observes, Australia is not only one of the highest per capita emitters of greenhouse gases, with a moral responsibility to act, but in terms of the potential impacts of climate change it is one of the most vulnerable developed countries. He notes that, as the lowest latitude developed country, Australia has a higher vulnerability to warming than other developed countries, and that: it will experience reductions in rainfall across populated regions; its water resources are already greatly stressed; its plants, animals and habitat have a low tolerance to temperature changes; it will have a stronger exposure to counter-adaptive trends such as cyclones and storm surges; and the frequency and severity of major droughts which the country already experiences will increase (Pittock 2009, p. 284–6).

Given this vulnerability, economists claim that it makes sense to act on climate change because the mitigation of greenhouse gas emissions will be far cheaper than adapting to the disproportionate effects of climate change. In 2008, Garnaut (2010, p. 12) advised the Prime Minister and the state premiers that effective mitigation would cost only 'one to two tenths of a percentage point per annum from the growth in Australian living standards over the next four decades. After that, the gains from avoided climate change are likely to outweigh additional costs of mitigation'. The message that it is economically imprudent to delay the cost of action was well received by the Labor Government, and has been its mantra since 2008. With so much emphasis placed upon Labor's ETS proposal, however, which has since imploded so badly, cost the Prime Minister his job, and lost the Government its majority, beyond its renewable energy target the Labor Government had no mitigation mechanism left to offer other than the failed ETS (AAP 2010a; Kelly 2010). For Labor, climate change policy has proven, as Garnaut (2008, p. xviii) anticipated, a diabolical policy challenge, and very much harder than any other public policy issue in living memory.

A review of constraining and facilitative factors involved in responding to the so-called diabolical challenge of climate action provides insight into the ETS's policy promise and messy political demise. The approach taken here is to place Australia's belated ratification of the Kyoto protocol and its proposed ETS pathway to emissions reduction within the context of its domestic politics. It follows Harrison and Sundstrom's (2010, p. 2) reasoning that:

> *The decisions regarding whether or not to ratify international agreements and to adopt national policies to mitigate climate change are in the end*

> *domestic political decisions, taken in the context of home-grown interests, national discourses, and domestic political decisions.*

These authors find, for example, that the cost of compliance can deter ratification of the Kyoto Protocol, but that a surge in the salience of climate change action can prompt a normative commitment sufficient to outweigh political self-interest. Arguably, this explains the Rudd Government's ratification of the Kyoto Protocol, and perhaps also the fate of its ETS, which was well supported by the Australian public in 2007, but increasingly less so as the potential cost of the scheme was debated. Harrison and Sundstrom (2010, p. 3) propose a focus on domestic self-interest, ideas and political institutions as a means of explaining both international positioning on climate change and the scope and efficacy of domestic climate policies. These factors discussed below will provide some insight into policy failure here, following discussion of the ETS.

Self-interest

The self-interest perspective goes to the heart of politics and is powerfully pragmatic. It sees politicians seeking to retain office and to pursue re-election, thus responding as far as required to public demands for climate change action whilst mindful also of the public's actual willingness to pay. The voices of interest groups are important influences, as are the relative level of lobbying influence and the salience of the climate change issue with the public at large. On the other hand, self-interest in a political/economic sense accounts for the degree of emissions intensity that needs to be dealt with, the political power of the fossil fuel industries, and the costs of compliance (Harrison & Sundstrom 2010, pp. 8–14). Self-interested behaviour, in economic and political terms, would explain the Howard Government's 'faking' of climate action – by failing to ratify Kyoto or to address Australia's emissions intensity, but saving face with the public by behaving as if its other climate change actions were world leading (Crowley 2007).

Ideas

The ideas perspective is highly significant but perhaps a less powerful frame. Harrison and Sundstrom (2010, pp. 14–16) see ideas playing out in several dimensions in terms of their impact on domestic climate change decision-making and policy design, for example in explaining political preference: parties of the left are typically more interested in regulatory or tax interventions, while parties of the right are more partial to voluntary mechanisms. This may not be the case in practice of course. The Australian Labor Party since the early 1980s has been highly partial to market-based policy mechanisms, with voluntary climate change action equally favoured by Labor and Coalition Governments in the 1990s. And, by contrast, in 2007 both the Coalition Government and the Labor opposition had reached the position of advocating the investigation of an ETS as the best potential mitigation mechanism. Thus norms *should* influence action, but they may not. A self-proclaimed climate change sceptic such as Prime Minister John Howard could, for instance, eventually accede to the public demand for an ETS whilst not believing at all in its efficacy.

Institutions

The institutional frame is significant but difficult to apply in isolation from political and normative influence. The most critical institutional characteristic defined by Harrison and Sundstrom (2010, pp. 16–18) is the electoral system, for example, the effect of a simple first-past-the post or a proportional electoral system, with the latter more readily facilitating the emergence of green parties. This has been the case since 1996 with the Australian Greens emerging nationally in the Senate to the point now of pressing the Gillard Labor minority Government for urgent and fundamental climate change action. Concentration of authority is also a critical institutional characteristic. When concentrated at the national level, for example, this may facilitate either leadership on climate action or, in Australia's case, recalcitrance by a powerful leader, for example, Prime Minister John Howard. On the other hand, diffusion of authority may provide multiple veto points, blocking change or facilitating policy innovation and transfer. Federalism also presents a vertical diffusion that may either obstruct national action, say by a resource-rich state attempting to resist carbon pricing, or facilitate innovative sub-national policy-making that may then 'race to the top'.

If we employ self-interest, ideas and institutions to analyse the climate change policy of the Howard Government, we can begin to illuminate the paradox of its refusal to ratify the Kyoto Protocol whilst embracing and pursuing its Kyoto targets. Normatively sceptical about climate science and impacts, politically self-interested in long-term office, captive to industry influence, and with a strongly concentrated authority and power base, this Government ranked jobs, industry and the economy above the environment. On the basis of this limited logic, it refused to ratify the protocol. However, consistently strong public opinion supporting ratification, and the need therefore to feed growing public expectations and alarm, prompted palliative action and the rhetoric that Australia was meeting its Kyoto target with world-leading actions. In fact, this target was an eight per cent increase in emissions, and its actions were largely to reduce land clearing (Crowley 2010). The legacy of this has been to make effective abatement all the more difficult moving into the post-Kyoto era. The struggle to reframe the climate change paradigm from symbolic to substantive action by the Rudd Labor Government, and to introduce an ETS, as described below, illustrates this difficulty very clearly.

ETS policy efforts

Emissions trading

By the 2007 federal election, Australia had very slowly come around to a position where both major parties supported emissions trading. The idea of an ETS had been around for some time, with a basic framework recommended to the Howard

Government in 1997 (Cornwell, Travis & Gunasekera 1997). An ETS would apply a market-based trading solution to the problem of emissions reduction. The Government would set an emissions reduction cap, issue permits to emit which the market would price, penalise non-compliance with emissions limits, and allow the trading of permits freely between market players. This could be a more flexible mitigation measure than a carbon tax, because it would create a market with incentives for reducing emissions, and also because it could articulate at an international level. The implementation of an ETS would require the measuring of emissions, the allocation of permits, the management of the yearly flow of permits, and the independent monitoring and verification of emissions reductions (Gans & Quiggin 2007). An ETS would be very unlikely to stand alone as a policy measure, however, and would be complemented, as it is in the UK, by other measures as required such as taxation and regulation as price-setting tools (DEFRA 2007, p. 22; NETT 2006, pp. 203–216)[2].

The Rudd Government noted the variety of ETS schemes then in operation: in the EU covering the energy and industrial sectors and linked to several non-EU countries (from 2005); in New Zealand covering all sectors and all emissions with phasing in (from 2008); in Japan as a limited trial scheme (from 2005); and in Canada covering half the country's emissions from electricity, oil, gas, iron, steel, cement, chemicals and fertiliser (from 2010). In the US, a number of regional trading initiatives had also sprung up: the Regional Greenhouse Gas Initiative covering power plants initially (from 2009); the Western Climate Initiative including western states from the United States (US) and provinces from Canada (from 2010); and the Mid-Western Greenhouse Gas Accord agreement for reducing emissions by six US states and one Canadian province (Australian Government 2011). Australia is also home to one of the world's first mandatory trading schemes – the sub-national New South Wales Government's 2003–2012 Greenhouse Gas Reduction Scheme, which has reduced emissions by 90 million tonnes (Garnaut 2008a, p. 356; Lyster 2004).

Australian efforts

In January 2004, the Australian states and territories established a working group to consider the design of a national ETS, publishing a discussion paper in 2006 and recommending a start date of 2010 (NETT 2006). This was important both politically, for signalling that the inaction of the Howard Government would not be tolerated by the states and territories, and for anticipating the actual design and implementation of a scaled up national ETS. Importantly as well, the states and territories agreed that a national ETS and complementary measures should cut emissions by 60 per cent by 2050. This then became a sub-national reduction benchmark that translated back into the respective state and territory policy environments. The Howard Government rejected the idea of emissions trading,

2 Milne (2008, p. 16) notes that in Europe, for example, 'which started with a terribly weak emissions trading scheme, it has been complementary measures such as the gross feed-in tariff for renewable energy that have driven the biggest CO_2 reductions to date'.

arguing that there were better ways to reduce emissions, such as capturing carbon dioxide and storing it underground, and supporting new technology including renewables (ABC Online 2006). However, it undertook its own review through a Prime Ministerial Taskforce on Emissions Trading, and by the 2007 election campaign the Howard Government, with the support of the Business Council of Australia and the Labor opposition, had at last committed to an ETS.

Carbon pricing through an ETS had long been recommended in Australian Government reports, and was seen as critical to stimulating the restructuring of the Australian economy by factoring carbon costs into decision-making (Daley & Edis 2010, p. 4). Without this pricing, or any regulations to restrict emissions, there would be no incentive to reduce emissions, nor to restructure operations to achieve efficiencies in terms of avoiding emissions. Furthermore, an ETS could align with the Howard Government's principle of 'no regrets' climate policy by reducing greenhouse gas emissions for no net cost, and indeed it could lower the cost of reducing emissions by 50 per cent (Cornwell, Travis & Gunasekera 1997). The publication of the Garnaut (2008) review was a significant agenda-setting milestone for its ETS proposal, notwithstanding criticism of its failure to recommend steeper emissions cuts[3]. The review was used in principle as the trajectory for the Labor Government's ETS, the Carbon Pollution Reduction Scheme (CPRS). However, the proposed CPRS was to succumb to political pressures all round, from vested interests and trade-exposed emissions-intensive industries, sufficient to distort it to the point of meaninglessness (Garnaut 2008a, pp. 314–8).

Labor's CPRS

The Government released a green discussion paper in July 2008 proposing an ETS, a White Paper proposal in December 2008, and the CPRS Bill into the parliament in May 2009. Its green paper was highly anticipated and acknowledged that Australia couldn't continue to pour carbon pollution into the atmosphere as if there were no social and environmental cost involved. This provoked a furore amongst emissions-intensive trade-exposed industries, including the aluminium, liquefied natural gas, coal, plastics, chemicals and paper sectors, with calls for compensation and a fixed price on emissions. Undeterred, following the release of the Garnaut review, the Government's White Paper, *Australia's Low Pollution Future* (DCC 2008), announced detailed plans for the introduction of a wide-ranging cap and trade scheme, the CPRS, with a 5 to 15 per cent emissions reduction target by 2020 – depending on the action of other nations. The CPRS was scheduled to commence in July 2010. It was designed to see emissions peak in its first year of operations, beginning with relatively mild emissions reductions and a 'modest starting carbon price estimated at $25 per tonne of emissions' (Stutchbury 2008). From the moment of both the green and the White Paper's release, every aspect of the proposed CPRS – its reduction target, its design and in particular its proposed compensation levels – was attacked from all quarters, from

[3] See Christoff (2010) for the gulf between climate science and Garnaut's recommendations.

green, industry and consumer lobbies, and by those both for and against climate action.

Whilst climate change decision-making has always been more politics than policy in Australia, this became particularly so in the wake of the CPRS green paper consultation process and White Paper proposal. Garnaut warned what could happen, in particular, in terms of industry manipulation of the proposed scheme. The Government could give 'in to well-organised interests by adopting weaker positions on the basic design of the scheme' that could 'undermine the policy objective of reducing emissions'. If policy makers could not resist pressure from vested interests, he argued, they should abandon the ETS and adopt a carbon tax[4] (Garnaut 2008a: 314 to 318). The overwhelming logic coming from Garnaut, the NETT and others was that delay of either ETS or taxation-based carbon pricing would greatly increase the cost of mitigating action later on. As Kelly noted, 'There is no room for playing politics with mitigation policy [and] thereby lifting its cost to the Australian economy' (Kelly 2008, p. 14). However, the overtly displayed pressure from emissions-intensive trade-exposed industries on the newly-elected Labor Government led both to a delay to any ETS or CPRS and to calls for all manner of exemptions to greatly displace costs away from the polluters.

The critical response

Proponents of climate action condemned the distorting complexities that were written by the Government into the proposed CPRS mechanism, as it attempted to appease industry groups after their virulent attacks upon its White Paper. A group of Australian economists concluded that the proposed modified scheme was inconsistent with the Government's goal of a global agreement to stabilise the climate. Further, the five per cent reduction below 2000 levels by 2020 (or fifteen per cent below with sufficient international action) was inadequate, with a more ambitious target needed to provide certainty and stimulate investment in renewable energy. The compensation proposal to energy-intensive trade-exposed industries would encourage expanded emissions, enhance the market power of such industries, and consolidate rather than reform the energy-intensive character of Australian industry. Neither was there any incentive for the voluntary reduction of emissions by householders, who would simply free up extra permits for energy-intensive industries to expand their pollution. The economists recommended a 25 to 40 per cent reduction target by 2020, the abolition of free permits to the biggest polluters, and a process for ensuring individual actions resulted in lower emissions not lower carbon prices (Stilwell 2009).

Garnaut himself expressed disappointment with the fate of the CPRS proposal. He said he was concerned that 'vested interests have had an undue influence on the scheme's design', saying that 'vested interests seeking to influence the process

---

[4] This is a course of action now being pursued by the Gillard Labor government by agreement with the Greens.

is not illegitimate, but it can lead to poor policy outcomes'. He was also concerned that the Government's emissions reduction target of five per cent was too low, with his recommendation being that up to 25 per cent was needed – in the context, however, of a global agreement. And he was concerned that the use of free permits as compensation to the highest polluters had gone too far and was too much of a reward following pressure from vested interests. In terms of the timing of the introduction of an emissions trading scheme, Garnaut (2009) suggested that mid-2010 would have been ideal. Hardships caused by the effects of global recession would then have been offset by investment in the structural change of Australia's economy that would tackle climate change and build economic resilience.

Kelly observed that climate change policy, the dream political issue for the Rudd Government at the 2007 election – indeed the greatest moral challenge of our time – had quickly transformed into a political and policy nightmare. The Government was 'trapped between the political virtue of tackling climate change, which it has exploited to the hilt, and the reality that meaningful action will hurt consumers and Australian companies and diminish Australia's growth and prosperity' (Kelly 2008). It was inevitable that the policy would be dragged down by politics and that the influence of affected industries would bite, with the result that the CPRS would become distorted and meaningless. Politics saw the logic of capping emissions, pricing carbon and stimulating restructuring as the quickest, cheapest means of reducing Australia's emissions profile all but abandoned. First the Labor Government attempted to 'walk back' its commitments, before losing control of the debate by failing to gain sufficient political support in the Senate; and then it abandoned the CPRS under pressure, for which it ending up paying the political price at the 2010 federal election.

ETS politics

The walk back

The 'walk back' is a colloquial political term that refers to a backing away from promises made by politicians, often in an election campaign, once the reality of the consequences sets in afterwards. And so, in a bid to appease industry and with an eye to achieving Senate approval, in May 2009 the Government announced: a delay to its CPRS by one year until July 2011, a fixed carbon price of $10 a tonne, and unlimited free permits to be sold to emitters until July 2012. Furthermore, for those industries already deemed eligible for free permits for up to 60 per cent of their emissions, an additional 10 per cent of free permits was added. For those industries deemed eligible for free permits up to 90 per cent, an additional five per cent of free permits was added. And in a bid to appease advocates of climate action, the Government announced a steepening of its potential reduction target from 15 per cent to 25 per cent below 2000 levels by 2020. But it would require global agreements of 15 to 25 per cent cuts for Australia to agree to a 15 per cent cut, and of greater than 25 per cent for Australia to adopt 25 per cent. Otherwise, Australia's reduction target would remain at five per cent.

The Opposition adopted this five per cent as well, although under their new leader, climate sceptic Tony Abbott, elected in November 2009, this was not to be achieved by any emissions trading or carbon taxation but rather by 'direct action' policy expenditure. The ousting of Malcolm Turnbull as Opposition Leader, because of his support for strong climate action and an ETS, was a dire development for the Rudd Government.

Just after the Labor Government's election in 2007, the international community met in Bali and agreed, in principle only, with the IPCC's recommendations that 25 to 40 per cent global emissions cuts were needed to avert dangerous climate change. So, the five per cent Australian commitment by the Rudd Government in 2009 dismayed the advocates of strong action. Furthermore, the Government observed that, with a CPRS in action, the five per cent cuts could be wholly achieved through the purchase of international credits – such as avoided deforestation credits – using CPRS revenue. A coalition of sixty community climate change groups wrote an open letter to Climate Change Minister Penny Wong condemning the target as appallingly low. They claimed it would undermine the UN meeting of the conference of parties to the Kyoto Protocol in Copenhagen in December 2009, which would be seeking to confirm the steeper reduction target through a formal, negotiated agreement. To appease other concerns, e.g. that householder action would simply free up permits to pollute, the Government proposed an Australian Carbon Trust that would formally recognise voluntary household action, with CPRS permits cancelled in proportion to emissions saved in energy efficiency measures (Carbon Positive 2009). Notwithstanding the walk back by the Government on its emissions reduction commitments, the CPRS would still have covered the top 1,000 emitting companies and have been applied across the economy, capturing 70 per cent of Australia's emissions. By comparison, the EU ETS covers only 45 per cent of their economy.

Losing control

Following the release of its CPRS White Paper, the Government had been caught between the pressures for climate action and the realities of the cost to industry, but also to consumers if the costs to emissions-intensive trade-exposed industry in particular were largely displaced onto *them*. As it walked back its previous commitments and moved towards tabling its watered down CPRS legislation, the Government encountered stiff political opposition from the Senate where it had no majority, and it was clear that its legislation would not be passed. The Prime Minister was intent upon having the CPRS in place by the December 2009 UN climate change meeting in Copenhagen; however, the Senate delayed the CPRS vote, and with the Government seven seats short of a majority, a stand-off ensued. The Coalition Opposition was opposed to any ETS before seeing what the US Waxman-Markey Bill was able to achieve, if anything, and before seeing the outcome at Copenhagen. It also embarked upon its own costing of the CPRS and a report into alternate carbon pricing options. However, by November 2009 the Government had walked its proposal back sufficiently to gain Opposition Leader

Malcolm Turnbull's support, reducing the support to households and increasing the assistance to industry. However, with his own party room's subsequent rejection of the CPRS by 41–32, Turnbull's support was to precipitate his own political downfall.

Throughout 2009, the Government had increasingly lost control of the CPRS debate and legislative progress once the CPRS Bill had passed the House of Representatives, particularly so when talk turned to the possibility that, if twice defeated in the Senate, the Government would threaten to call a double dissolution election in a bid to force the Senate's hand. The Government gambled that Turnbull would not want to fight an early climate change election and would gain his party's support for an amended CPRS simply on these grounds alone. However, with public opinion well set against an early election, the impending threat of a double dissolution election was seen as increasingly hollow. Events took a turn for the worse for the Government when Turnbull was deposed and succeeded by a climate change sceptic Tony Abbott. Under Abbott, the Coalition would only favour an ETS scheme that issued free permits up to the level of 100 per cent for the highest emitters. Meanwhile, the 'great big tax' rhetoric that Abbott peddled about the CPRS resounded with the Australian public, leaving the CPRS unable to make traction when pitched against concern over economic recovery. An election in such circumstances, with a new Opposition Leader enjoying a political honeymoon period, was considered ill-advised (Shanahan 2009).

Abandonment under pressure

Within the space of twelve months over 2009–2010, climate change would see both an Opposition Leader (Turnbull) and a Prime Minister (Rudd) deposed, a surge in the Green vote at the 2010 federal election, and a hung federal parliament as the election outcome. Labor was reduced to minority Government status and only returned to power by negotiating with the Greens and a collection of rural independents. These are huge political stakes, bearing in mind that the 2007 election result that swept the Coalition Howard Government from power was largely regarded as Australia's first climate change election (Rootes 2008). From the Government's rhetorical high, advocating urgent climate action in 2007, over a period of only two years its policy was completely subsumed by politics as it abandoned plank after plank of its increasingly compromised CPRS. After only twelve months, the compromise was very clear: 'All up, the Government is offering so much protection [to polluting industries] that it risks defeating the purpose of the ETS: to encourage a drastic change in behaviour and to adapt to a much lower use of carbon. It may be clever politics but at the expense of good policy' (Steketee 2008). Twelve months later, Ross Garnaut commented that the 'whole process of policy-making over the ETS has been one of the worst examples of policy-making we have seen on major issues in Australia. I suppose it was never going to be easy, but the way it has broken down was extraordinary' (ABC 2009).

The CPRS fell into a void following the disappointing 2009 Copenhagen meeting and into the 2010, with the Government dropping all talk of a double dissolution election even after a new CPRS bill was again blocked in the Senate. In April, Rudd announced a 'credibility snapping' decision to delay the CPRS until 2013, effectively abandoning climate policy in the lead up to the 2010 federal election. He was severely pilloried in the press for turning his back on, in his own words, 'the greatest moral, economic and environmental challenge of our time'. Van Onselen argued that Rudd would take a less crystallised climate policy to the 2010 election than John Howard had taken to the 2007, and that the delay of the CPRS was also in keeping with Opposition policy. Rudd blamed the Opposition's obstructionism - and also the need to see what the rest of the world would decide in terms of climate targets and mitigation mechanisms - for his decision to delay, which again mirrored Opposition logic. Rudd was caught out contradicting his previous rhetoric at every turn, including where he had claimed that to delay any longer would be reckless and irresponsible for our economy and our environment. And finally, he was broadly condemned for lacking the political courage to take on a new Opposition leader in a climate change election (Van Onselen 2010).

The political toll

Prime Minister Rudd was to pay the ultimate political price for his abandonment of climate policy, but by early 2010 there were also question marks over bungled climate programs administered by the departments of environment and climate change. Programs such as the Solar Rebate, Green Loans and Housing Insulation subsidy schemes unravelled under Labor Environment Minister Peter Garrett (Berkovic 2010). Allegations ranged from rorting and shoddy services by contractors, to massive cost blowouts and poor management. The success of the twenty per cent Renewable Energy Target, which had been separated out from the CPRS Bill and passed with Opposition support, was undermined by problems with implementation that needed a distinction to be introduced between household and commercial schemes. Rudd transferred some of these problematic programs to the Department of Climate Change, which now had no CPRS to administer, adding energy efficiency to this department and creating a new energy efficiency ministry. The impression he created, however, was very much indecisive, back-footed and back-flipping, and this contributed strongly to both the Labor Government's and Rudd's own steep slump in the polls, with the Coalition leading 43 to 35 per cent by May. There was also an implosion in support for Labor's handling of climate policy to level-pegging with the Opposition (see Table 3.1).

Prime Minister Rudd appeared to have lost his way and his confidence after the failure of the Copenhagen meeting to resolve climate targets. The early signalling of his own Government's mediocre emissions reduction target was indeed taken by critics to have undermined global negotiations for a post-Kyoto agreement in Copenhagen in December 2009. Cohen observed that 'Kevin Rudd appeared bewildered as his plan to make Copenhagen his personal triumph dissolved before his eyes' (Cohen 2009). Climate policy failure and the shelving of the ETS stripped the Government of its policy credentials, possibly even more so for failing to live up to the Prime Minister's own bold and very welcomed pre-election promises in 2007. After poor polling in May 2010, Deputy Leader Julia Gillard asked the Prime Minister for a leadership vote, but she was elected unopposed as Kevin Rudd realised he did not have the numbers to continue. Having made history by deposing a first term Prime Minister, it took the new Prime Minister a couple of weeks to get to climate policy and her announcement of a citizen's assembly to decide the issue. This was possibly the greatest misjudgement on climate change action made by any Australian Prime Minister, and was broadly condemned as the electorate headed for the polls on 21 August. The political toll of climate policy failure included the Government's loss of its majority in the 2010 election that saw an historic swing to independents and the Greens.

Table 3.1 Party best able to handle climate change

% Support	Jul 2008	Oct 2008	Feb 2009	Feb 2010	Apr 2010
Labor Party	45	37	38	35	27
Liberals/Nationals Coalition	18	22	19	30	26
Someone Else	17	20	23	17	26
None	4	5	4	4	8
Uncommitted	16	16	16	14	13

Source: Newspoll Market and Social Research, *The Australian,* 4 May 2010.

Question: Who do you think would best handle the issue of climate change?

Conclusions

How do self-interest, ideas and institutions help to analyse the politics of climate change policy failure and the thwarted ETS under the Labor Government from 2007 to 2010?

Political self-interest is a powerful explanation that can be read in many ways, with Kevin Rudd championing the ratification of the Kyoto protocol from opposition and riding high on the public concern for action in the 2007 election campaign. Once in office as Prime Minister, he was able to buy time waiting for the Garnaut review, which was released in 2008, but already the public was becoming uneasy about his refusal to consider the 25 to 40 per cent targets proposed in Bali in 2007. But the costs of compliance were sinking in with the Labor Government, and the pressure was mounting from the likely affected industries, consumers and their respective organisations. There was no doubting public support for action, however, with the Climate Institute releasing an opinion poll, timed with the release of the CPRS Bill, showing that 78 per cent of

Australians were at the time concerned about climate change (Climate Institute 2009). The Government's own research confirmed this, but this did not prevent the Government balking at implementation of its own agenda, or what Michael Steketee called Labor's 'great climate change choke' (Steketee 2010).

It is difficult to see normative perspectives as in any way as significant as material interests, both political and economic, in terms of the politics of climate change policy failure and the thwarted ETS. Perhaps the most compelling display of both normative behaviour and lack of political self-interest was Opposition Leader Malcolm Turnbull's dogged persistence in supporting an ETS, even knowing this would cost him his leadership. Still, he stuck by his 'political principles' after losing the leadership, claiming that his views were unchanged despite contradicting his new leader's views. He added: 'The Coalition has a policy on climate change....it's not ideal from my point of view...but it does stand in stark contrast to the Labor Government which has no policy at all' (AAP 2010b). In terms of normative debate about climate science, the departure of John Howard, the leadership of Malcolm Turnbull, and the Garnaut report were all factors in neutralising climate change scepticism and focusing attention on mitigation. However, normative debate about market-based versus other mitigation mechanisms took a strange turn with Tony Abbott as Coalition leader opposing a market-based approach, but also with Labor under new Prime Minister Julia Gillard opposing a carbon tax. It seems normative climate change debate in Australia remains ever captive to political and economic self-interest.

There is interplay between normative debate, political self-interest and institutional capacity if we consider the attempts by the Australian Greens to gain traction on climate change in the Australian Senate. The Greens did not support the CPRS Bill, claiming that the Greens cannot and will not support a scheme that is environmentally ineffective and economically inefficient', and that it is difficult to see how anyone serious about climate change could support the Government's proposal. And, the Greens held that the Government's crash through or crash approach to the ETS, with the convoluted and intricate arrangements for compensation, offsets and rebates - muting the price signal and undermining the purpose of the ETS exercise - was hardly the leadership that Kevin Rudd had promised on climate change (Milne 2009, p.6). In practice, the fact that the electoral system had delivered Greens into the Senate did nothing for their influence on the shape of the ETS, as the Rudd Government refused to discuss it with them[5]. Indeed, his remote, obstinate and autocratic behaviour, coupled with his concentration of the Government's power very narrowly in a team of four[6], contributed to his own downfall.

In the end, the most compelling explanation of the politics of Australian climate change policy failure and the thwarted ETS remains the structure of its natural resource-based, carbon-intensive economy and the influential lobbying of the

---

5 This was set to change in July 2011 when the Greens assumed the balance of power in the Senate.

6 The Prime Minister, the Deputy Prime Minister, the Treasurer and the Finance Minister.

industries likely to be affected by mitigation. This in turn is significant in generating political and economic self-interest, leading politicians to brave the tide of public support for climate leadership by refusing to act for fear of the costs of compliance to themselves and to the country. Whilst popular and political commentators have long pointed to the constraining influence of a greenhouse mafia over Australian climate change policy, the influence became overtly apparent during the ETS debate. It was remarked upon by classical economist Ross Garnaut as the major derailing factor in the effort by the Rudd Government to put a price on carbon, along with the failure of Rudd's own leadership to withstand the influence of affected industries. For this, the Labor Government paid at the polls, and its next effort to price carbon begins under a new leader – but this time reliant upon the Greens and independents in the House of Representatives and the Senate to pass the legislation. It remains to be seen whether climate change will remain more politics than policy in these vastly changed circumstances. But that's another story…

Chapter 4

Lost energy policies, opportunities and practice

MARK DIESENDORF

Introduction

Two great threats, both deriving from fossil energy consumption, loom over human civilisation in the twenty-first century: climate change and loss of energy security.

Human-induced climate change arises primarily from the combustion of fossil fuels to generate electricity and heat and to run motor vehicles. Since industrialisation, Earth's carbon dioxide (CO_2) concentration has been increasing and in 2010 reached 390 parts per million (ppm). Yet for at least the 660,000 years before industrialisation it oscillated between 180 and 280 ppm under the action of natural forces. The increase in Earth's average temperature since industrialisation has been 0.76°C and the projected increase by 2100, under conservative assumptions, lies in the range 1.5 to 4.5°C. Projections of current trends in emissions lift this range to 2 to 7°C. The consequences already being observed are the melting of polar caps and glaciers, a steady rise in sea level, acidification of the oceans, enhanced loss of biodiversity, and increases in the prevalence and severity of heat waves, droughts, wildfires, floods and probably strong hurricanes (IPCC 2007; Allison *et al.* 2009).

Industrial society has been built upon abundant and hence cheap fossil fuels. Oil in particular is the principal fuel for transportation, petrochemicals, agriculture and food production. Yet, as the result of our insatiable demand and finite supplies, we have now reached the peak in global conventional oil production, or are very close to it (Aleklett *et al.* 2010; IEA 2010). New sources are increasingly difficult to access in terms of economic cost, energy inputs and environmental hazards, as witness the recent major oil spills in the Gulf of Mexico and off the

Western Australian coast. At current trends, peak oil could be followed by peak coal before 2050 (Mohr & Evans 2009).

Yet there is a common partial solution to both threats: making a transition to an ecologically sustainable energy system based primarily on the efficient use of renewable sources of energy (Diesendorf 2011a). Sustainable energy technologies have matured to the extent that it is now technologically feasible to consider realistic scenarios for national, regional and global energy systems that become based predominantly or entirely on renewable energy by 2050 (Sørensen & Meibom 2000; Makhijani 2007; Sørensen 2008; Mason, Page & Williamson 2010; Teske & Vincent 2008; EREC 2010; CAT 2010; Daley & Edis 2010; WWF 2011,).

While the costs of following these scenarios are still uncertain, it appears likely that the economic savings from using energy more efficiently could pay for a large part of the additional costs of renewable energy (McKinsey & Co 2009). Furthermore, the costs of continuing with business-as-usual are likely to be much greater than the costs of climate mitigation (Stern 2006).

This is the context in which this chapter examines and compares the energy policies of two Australian federal governments: the Howard Liberal-National Party Coalition government of 1996-2007 and the Rudd-Gillard Labor government from its election in November 2007 to May 2011 (when this chapter was finalised). The focus is on energy, especially electricity and heat, which are together responsible for about half of Australia's greenhouse gas emissions. Due to space constraints, energy efficiency and transport receive only passing mention, although they are both important, as are the two non-technological drivers of greenhouse gas emissions-growth, namely economic activity and population (Diesendorf 2009, pp. 119-124).

Under the twin threats of climate change and energy insecurity, an effective set of government policies would rapidly transform the Australian energy system into one that is based predominantly on the efficient use of renewable energy. Unfortunately, as shown in this chapter, neither government has implemented such policies. Instead, while paying lip service to the first of these threats and glossing over the second, both governments have enhanced the dominant role of fossil fuels and kept renewable energy in a niche role. In contrast to government failure, this chapter recommends effective policy options to create an ecologically sustainable, socially just and low-risk energy future for Australia.

Energy policies of the Howard government

The Howard government's approach and priorities in energy policy were expressed in its 2004 Energy White Paper:

> *Australia's access to low-cost energy also underpins much of our industrial base…Developing Australia's abundant low-cost energy resources is a key to our future prosperity…For the foreseeable future, coal, oil and gas will meet the bulk of the nation's energy needs with gas, wind solar and biomass*

> *increasing their share of Australia's energy supply (Australian Government 2004, Overview, pp. 1, 3, 5).*

As with renewable energy, the White Paper pays lip service to the threat of human-induced climate change, but states that Australia's greenhouse gas emissions 'are too small for Australia to make a difference on its own' (p. 24). It attempts to justify the government's refusal to ratify the Kyoto Protocol on the grounds that it is allegedly ineffective and 'does not include all of the largest emitters in the world', singling out developing countries for special mention (p. 24). Furthermore, it reassures the reader that energy security is a minor problem and states that pursuing alternative fuels now 'would weaken Australia's competitiveness and potentially weaken its energy security position' (p. 22), the latter being an extraordinary claim. The White Paper frames the government's modest policies to foster energy efficiency and renewable energy as taking 'practical action now to prepare our economy and society for the future' (p. 25), while implying that this future is far away.

Leaked documents revealed that behind the scenes the fossil fuels industry influenced strongly the White Paper through a group hand-picked by the government to form the Low Emissions Technology Advisory Committee. Fossil fuel producers—including Exxon Mobil, Rio Tinto and BHP Billiton—and big fossil fuel users and generators—Alcoa, Holden, Boral, Amcor, Energex, Edison Mission and Origin Energy—dominated the committee. There were no representatives from the renewable energy industry. Thus and by other means, the corporations with vested interests in fossil fuels worked closely with the government to develop the White Paper. The documents show that they actually plotted with the Prime Minister and industry minister to hold back wind power and to avoid or delay the expansion of the Mandatory Renewable Energy Target (ABC 2004; Miller 2004; Hamilton 2007). Subsequently, Guy Pearse, a former adviser to a minister in the Coalition government, confirmed that the big greenhouse gas-emitting industries, calling themselves the 'greenhouse mafia', had boasted to him that they wrote government policy on climate and energy (Pearse 2007).

In response to public concerns about climate change, the principal policy programs of the Howard government announced in the White Paper were the Low Emissions Technology Demonstration Fund (LETDF), the Renewable Energy Development Initiative (REDI) and Solar Cities. These and other renewable energy policies of the Howard government are summarised in Table 4.1. In practice, LETDF only funded grants for five projects during its short lifetime. Four of these were for fossil fuel projects and the remaining project, a subsidy for the solar photovoltaic power station proposed by Solar Systems for Mildura, never received its allocated $75 million. The company was placed in administration in 2009 and was taken over by Silex Systems Ltd in 2010.

Solar Cities was a politically astute program because it set up energy efficiency and renewable energy programs or 'trials' in initially five locations around Australia, creating the impression of extensive activity. In practice, the implementation only affected a small part of each 'city', and the technologies (apart from smart meters) were commercially available and did not need trials. Solar Cities is a token small-scale substitute for what is really needed, that is, the powering of all of Australia's towns by renewable energy used efficiently.

REDI made some useful contributions in the industrial sector. However, the most effective program of the Howard government was the Mandatory Renewable Energy Target (MRET), which commenced in 2001 and provided a quasi-market mechanism for the expansion of the demand for and supply of the lowest-priced renewable energy sources. Despite the tiny magnitude of the target (9500 gigawatt-hours per year) and several other flaws, it fostered growth mainly in hydro-electricity, solar hot water and wind power. The Howard government rejected the recommendation of its own Tambling (2003) committee to expand and extend the MRET scheme. The result was that a boom in wind power lasted until 2006, when the installed capacity to reach the tiny 2010 target was essentially reached. Then, the factories that had opened to build wind turbine components in Portland, Victoria and in Wynyard, Tasmania closed, and the wind industry remained in the doldrums until January 2011 when the Extended Renewable Energy Target (ERET) was finally implemented in an effective form - over three years after the Rudd government took office.

The photovoltaic (PV) rebate, initially $4,000 per system, and the solar hot water rebate, initially $1,000 per system, expanded the local market for residential solar systems. Although the PV Rebate Program was an expensive way of cutting greenhouse gas emissions, it provided, together with the Solar Hot Water Rebate program, a means for households to become involved directly in renewable energy use. BP Solar opened a manufacturing plant for solar PV modules in Sydney, employing 300 people. Solar hot water manufacturers, mostly based in Perth, thrived.

Another early policy of the Howard years was the Greenhouse Gas Abatement Program (GGAP) which provided nominally $400 million over four years, commencing in 2000. Because it avoided funding renewable energy and efficient energy use, GGAP was described by the Australian Greens as a 'corporate welfare program' for the big greenhouse polluters (Brown 2001).

With the support of the Labor Opposition, the Howard government expanded uranium mining and export. It also commenced the groundwork for nuclear energy by holding a Uranium Mining, Processing and Nuclear Energy Review, supervised by a committee composed mostly of people from the nuclear industry or nuclear physics (Commonwealth of Australia 2006). Not surprisingly, the committee's report was very supportive of nuclear energy and attempted to put down renewable energy.

The Howard government funded four Cooperative Research Centres to support the coal industry: on black coal, brown coal, carbon capture and storage (CCS), and mining technologies (CRC 2011). It also funded the CSIRO Energy Transformed Flagship, with the lion's share going to fossil fuels, setting in place the foundations for research and development in CCS.

Thus, the Howard government kept renewable energy as a niche market and several of its Ministers even made public attacks on wind power, the least costly of the renewable electricity sources (Diesendorf 2007, pp. 107-109). In reaction to the strong public support for renewable energy, the government implemented some token policies (mentioned above) with minimal government funding. However, despite much government propaganda for fossil fuels (for example in the Energy White Paper), MRET received such a strong positive response from the young renewable energy industries, the public and the Tambling Review that the Rudd Labor Opposition made its expansion in size and extension in time the second most important feature (after CCS programs) of its energy policy for the 2007 election.

Energy policies of the Rudd-Gillard government

The Rudd government greatly expanded federal support of coal power with CCS, making it its top technology priority for greenhouse gas reduction. In addition to the four Cooperative Research Centres on coal, Rudd's CCS programs included:

- the CCS Flagships Program for large-scale demonstration plants ($2 billion);
- the National Low Emissions Coal Initiative, a research, development and demonstration (RD&D) program ($385 million);
- Australia's contribution to the Global Carbon Capture and Storage Institute ($100 million per year for an unspecified period); and
- an expansion of CSIRO's research program on CCS (budget unknown).

In addition, several states have committed hundreds of millions of dollars to CCS.

This level of support is remarkable, considering that CCS systems are technologically unproven and are unlikely to be able to make a significant contribution to Australia's energy supply before 2025 or 2030 (Saddler *et al.* 2004). The many scientific, technological and economic uncertainties of CCS tend to be glossed over by its proponents (Hansson & Bryngelsson 2009).

In contrast, several renewable energy technologies are either in commercial mass production globally (solar hot water; on-shore wind power; bio-electricity; bioenergy cogeneration; photovoltaics; and geothermal heat). Others (concentrated solar thermal electricity with thermal storage; off-shore wind) are pre-commercial and within five years of becoming ready for full-scale commercial mass production. Before the November 2007 federal election, Rudd Labor made several promises to support the growth of renewable energy in Australia, including: a substantial expansion and time extension of Howard's MRET into ERET; a Renewable Energy Fund of $500 million for demonstration and deployment; an Energy Innovation Fund of $150 million for research and development (R&D); and an increase in Howard's rebate for residential PV to $8,000 per system. According to an exit poll conducted at the 2007 election, Rudd won government partly because of these promises (Climate Institute 2007). Public support for climate action and renewable energy remained high (Auspoll 2008, p. 8; Climate Institute 2008, p. 7).

Nevertheless, the Rudd government terminated, unnecessarily delayed or weakened almost all of its promised programs for renewable energy and several of the modest Howard government policies as well. By June 2010, when Julia Gillard displaced Kevin Rudd as Prime Minister, only a small fraction of the promised funds had actually been spent. The fate of some key programs remains as follows:

- The Photovoltaic Rebate Program was means-tested from May 2008, then terminated in June 2009. Subsequently, under Rudd, it was replaced with the Solar Credits Scheme, which gives a smaller subsidy that is not paid by government but by electricity consumers.
- Howard's Remote Renewable Power Generation Program (RRPGP) and Cities for Climate Protection were both terminated. Although both schemes were tiny in terms of funding, they gave assistance to remote rural electricity users and local governments respectively, helping renewable energy to become more mainstream.
- Neither the Energy Innovation Fund nor the Renewable Energy Fund made any grants for renewable energy in the first two years of office of the Rudd government. (There was some funding for geothermal drilling, which strictly speaking is not renewable energy.) The former fund was rolled over into the Australian Solar Institute, which began to announce grants in December 2009, and the latter into Renewables Australia, renamed the Australian Centre for Renewable Energy (ACRE), which is an umbrella organisation responsible for managing several programs and measures. Most of ACRE's funding is allocated to the Renewable Energy Demonstration Program (REDP), which announced grants of $235m in May 2009 and $92m in May 2010.
- On two occasions between June 2010 and May 2011, the Gillard government withdrew funding that had been allocated previously to renewable energy programs. One announced cut, $220m from the Solar Flagships Program and $150m from the Solar Hot Water Rebate to pay for a proposed Cash for Clunkers program, was reduced under pressure from the Greens. Cuts of about $400m announced in the May 2011 federal budget include deferral of funding for the second stage of Solar Flagships and the stretching out over 14 years of the allocation of $100m to the Renewable Energy Venture Capital Fund – a cut in other words.

The fate of the promised ERET, the principal policy for building the market for renewable energy, deserves particular attention. It would have been a simple matter for the Rudd government to increase the target and timescale in the legislation for the existing Howard MRET within a few months of taking office. Instead, the new government delayed the process by passing it onto the Council of Australian Governments (COAG) and eventually put it to Parliament in August

2009, nearly two years after taking office. The scheme had been designed to fail. Most of the target would be taken up with residential hot water systems from solar energy and electric heat pumps and by 'phantom' Renewable Energy Certificates (RECs) granted to purchasers of residential solar PV systems under the Solar Credits Scheme. There would be no room for large-scale renewable energy systems (Buckman & Diesendorf 2010). As expected, the price of RECs collapsed soon after the scheme was implemented. Then, while workers were being laid off and several renewable energy power stations faced bankruptcy (Berkovic 2009; McCutcheon 2009), the government delayed action for several more months by holding another inquiry. Eventually, it revised the scheme, separating large-scale from small-scale renewable energy systems and then introducing another long delay until implementation in January 2011.

Table 4.2 is an attempt to summarise the purpose and fate of many Rudd government programs that purported to support renewable energy. The reader can verify by inspection the complexity of this system, where programs were frequently started and stopped, deferred, renamed, rolled into other programs, split and consolidated. Even experts in the field have difficulty in following the money trails. The net effect was that very little of the promised government funding had actually been spent by July 2010, when Julia Gillard took the Prime Minister's position from Kevin Rudd and initiated the 2010 federal election campaign. It appears that hot rock geothermal power may have spent the most government funding up to that time, and that funds being allocated to R&D in solar electricity may become the major expenditure in 2011—but it's still too early to tell. Neither of these areas of funding is an immediate threat to coal power, because geothermal power is not yet at the demonstration stage in Australia and solar R&D will take several years at least to be translated into market benefits.

Any remaining doubt that the Rudd government's energy policy was drafted by the greenhouse mafia was dispelled by the process it set in place for the development of its own energy white paper. The energy minister appointed a 'High Level Consultative Committee' which he described as 'one of the key sources of advice for the development of the White Paper', representing 'a cross-section of stakeholders in Australia's energy sector … from discovery and exploration through to export and end use, and in both existing and emerging technologies' (quoted by Keane 2009). The committee was dominated by representatives from: Shell; BHP's uranium division; Santos; Woodside; Rio Tinto; Origin; AGL; Xstrata; the Energy Supply Association of Australia; the Australian Petroleum Production and Exploration Association; and the Construction, Forestry, Mining, and Energy Union (DRET 2011). There was no one capable of representing renewable energy. After complaints from the renewable energy industry and some academics led to adverse media comment, the minister appointed a renewable energy practitioner from a company developing wave power. Since this is not yet a commercial technology, it is hardly likely to influence proceedings. There were no representatives for wind or solar energy. The situation is almost identical with the process to develop the Howard government's Energy White Paper, except that the current Consultative Committee is not secret.

The 'centrepiece' of the Rudd government's climate change policy was an emissions trading scheme (ETS), called by the government the Carbon Pollution Reduction Scheme (CPRS). An effective ETS would place a price on carbon emissions and make the polluter pay. Thus, greenhouse-intensive industries and products would become more expensive and a shift in the national economic structure toward less polluting technologies would be achieved. However, the design of the CPRS was so strongly influenced by the greenhouse mafia that it would have locked-in the business-as-usual scenario. Up to 2020, the biggest greenhouse polluters would have received free emission permits of approximately $22 billion in value (Daley & Edis 2010, table 2.1). The medium-sized polluters could have offset all their emissions in cheap overseas projects that may not have even produced emission reductions overseas. There was no mechanism to ensure that the CPRS would result in any emission reductions in Australia. It was rejected in Parliament by the Coalition as 'a great big new tax', and by the Greens on the grounds that it would have been ineffective. Faced with an election, Rudd and Gillard promised that it would not be revived before 2013, if ever. If it had been passed, the carbon price would have been too low for about twenty years to assist renewable electricity to compete with conventional coal power. At best, it would have sent a signal to potential investors in conventional coal power that they would be taking a financial risk to proceed. Gas power would have thrived while it lasted.

However, shortly after the 2010 federal election, when Gillard Labor only regained power with the support of the Greens, Gillard formed a Multi-Party Climate Change Committee to explore options for the introduction of a carbon price. The Committee's proposed two-stage plan starts with a fixed carbon price period for three to five years before transitioning to an emissions trading scheme. It could commence on 1 July 2012.

Energy security under the Rudd government received even less action than climate change. Following a 2007 election promise, the Rudd government commissioned a National Energy Security Assessment (Australian Government 2009). Despite soaring oil prices before the global financial crisis (GFC), the report manages to be optimistic. It doesn't even mention the term 'peak oil', only hinting at the future possibility of 'a tight supply/demand balance'. With great faith in the market, it reassures the reader that 'Australia's declining liquid fuel self-sufficiency does not necessarily imply reduced energy security'. It claims that, in 2018, the situation will be in broad terms the same as in 2009, with adequacy and reliability both classified as 'high' and affordability as 'moderate'. In reality, Australia's oil production peaked in 2000, and in 2009 it was importing 55 per cent of its consumption. By 2020, imports could reach 80 per cent of consumption and could cost several tens of billions per year (Diesendorf *et al.* 2010, p. 22).

Can the global market provide security and affordability? According to the International Energy Agency, global production of conventional oil peaked around 2006 (IEA 2010), although we cannot be sure until the GFC is over. On prices, the report's main concern seems to be that a carbon price might lift the prices of petrol and diesel. Actually, it would only be a few cents per litre. For comparison, the price of a barrel of crude oil tripled between 2005 and 2007-08, reaching US$140 per barrel. During the GFC it dropped to US$40 per barrel, but returned to over US$100 per barrel in 2011. Australia has no strategy to deal with the probable reduction in the world's oil supplies in the near future. Unlike the US, Australia has no strategic oil reserve. Unlike Europe, Australian towns and cities have either poor public transport systems or, in the case of Sydney and Melbourne, potentially good but neglected systems approaching full capacity. The remedies should be obvious (Diesendorf *et al.* 2010).

State and territory policies

The principal policy programs by Australian state and territory governments to expand renewable electricity are the feed-in tariffs listed in Table 4.3. Unfortunately most are net tariffs, meaning that they only pay a premium price on the difference between the quantity of renewable electricity fed by a household into the grid and electricity purchased from the grid by the household, provided the former is greater than the latter. They are also limited to small-scale RElec (renewable electricity) systems, except in the ACT. In July, 2010 the Victorian Labor government foreshadowed a feed-in tariff for large-scale systems, but it was unclear whether this would continue under the incoming Coalition state government.

Table 4.1 Renewable energy policies and programs under the Howard Coalition government, 1996-2007

Policy	Description	Comment
Low Emission Technology Demonstration Fund (LETDF)	Nominally $410M announced in 2004.	In practice only 5 projects awarded grants, of which 4 were for fossil fuel projects. Remaining project (solar) did not receive the allocated funds. In total only a fraction of the Fund was spent.
Mandatory Renewable Energy Target (MRET)	A tradable certificate scheme commenced 2001 & funded by electricity consumers. Target of additional 9500 GWh/yr from RE in 2010 and for a further decade.	Too successful. Government rejected recommendations to expand and extend it. Tiny target essentially achieved in 2006, leading to closure of factories manufacturing wind turbine components in Vic and Tas.
Renewable Energy Development Initiative (REDI)	From 2004, AusIndustry grants program from nominal fund of $100M to support the development of RE technology products, processes or services that have strong early stage commercialisation and emissions reduction potential.	$67M paid in grants to 29 projects. Program ended 28-4-2008.
Solar Cities	Funding of up to $94M for 5 parts of cities or regions to install commercial EE & RE technology and smart meters. Cities chosen by Howard govt: Adelaide, Alice Springs, Blacktown (suburb of Sydney), Central Victoria and Townsville.	A feel-good program that produced little new knowledge or renewable energy generation. A substitute for making all cities 'solar cities'.
Photovoltaic Rebate Program	Commenced 2000 with rebates of $4000 per system.	In 2007 both Labor & Coalition made 2007 pre-election promises to increase it to $8000 per system. For outcome, see Table 4.2.
Cities for Climate Protection (CCP)	Assisted local governments and their communities to reduce their greenhouse gas emissions.	Tiny funding of about $4M p.a. to cover about 800 local governments. Despite tiny grants, taken up with enthusiasm by local governments.
Remote Renewable Power Generation Program (RRPGP)	Grants for small-scale RElec systems in remote areas.	Substituted for diesel, not coal.
Solar Hot Water Rebate	Announced July 2007. Program to assist home-owners to replace electric storage hot water systems with solar or heat pump systems. Initially $1000 rebate	
Wind Energy Forecasting Initiative	From 2004, provided up to $14M for development and installation of software and systems for forecasting wind energy.	
Renewable Energy Equity Fund (REEF)	From 1999 $17.7M for venture capital, administered by Innovation Australia.	

Notes: This table is not comprehensive. RE = renewable energy; RElec = renewable electricity; EE = energy efficiency.

Websites (accessed 24/5/2011): CCP <http://www.environment.gov.au/archive/settlements/local/ccp>; RRPGP <http://www.climatechange.gov.au/what-you-need-to-know/renewable-energy/rrpg-program.aspx>; Solar hot water rebate <http://www.climatechange.gov.au/government/programs-and-rebates/solar-hot-water.aspx>; Solar Cities <http://www.climatechange.gov.au/government/programs-and-rebates/solar-cities.aspx>. LETDF: <http://www.ret.gov.au/energy/clean/cei/low_emissions_technology_demonstration_fund/Pages/LowEmissionsTechnologyDemonstrationFund.aspx.

Table 4.2 Renewable energy programs under Rudd-Gillard Labor government, Dec. 2007-July 2010

Policy/program	**Description**	**Comment**
Expanded Renewable Energy Target (ERET)	Certificate scheme funded by electricity consumers except certain energy intensive industries. Target of 45,000 GWh per year (including previous MRET target of 9500 GWh per year) in each year from 2020–2030.	Pre-election promise in 2007; Labor government delayed implementation until August 2009, when design flaws caused collapse of certificate price; implementation of revised ERET with separation of small- & large-scale systems on 1-1-2011. Since then large-scale target has assisted mainly wind, while small-scale scheme has assisted mainly PV.
Renewable Energy Fund	$500M for grants for dissemination of RE. Comprised a nominal $300M Renewable Energy Demonstration Program, $50M Geothermal Drilling Program and $15M 2nd generation Biofuels Program.	Pre-election promise in 2007. Apart from geothermal drilling, which was rolled over into a separate fund (see below), no funds were allocated in 2008. All subprograms rolled over into the new Renewables Australia/ACRE (see below) in May 2009 and Renewable Energy Fund was closed.
Renewable Energy Demonstration Program (REDP)	Initially $300M subprogram of Renewable Energy Fund, aiming to demonstrate RE technologies with grants in range $50–100M ($1 from govt for every $2 from private sector).	In May 2009 budget, a total of $235M of REDP grants was allocated and in November 2009 was awarded to 4 projects: two geothermal, one wave and one mixed RE. The solar part of REDP ($135M) was rolled into Solar Flagships Program (see below) and $365M into new body Renewables Australia/ACRE (see below). In May 2010 REDP grants of $92M (with conditions) announced for 2 solar thermal power projects.
Geothermal Drilling Program	$50M subprogram of Renewable Energy Fund. Grants for development, commercialisation & deployment.	About $50M allocated in 2 rounds. Rolled into Renewables Australia/ACRE in May 2009.
2nd generation biofuels R&D	Total of $15M for grants for R, D&D	Rolled into Renewables Australia/ACRE in May 2009.
Energy Innovation Fund R&D	Originally $150M for R&D grants in RE	2007 pre-election promise. No expenditure in 2008 or 2009. Then $100M of fund allocated to support Australian Solar Institute (see below). Fate of remaining $50M of allocation unclear to this author.
Australian Solar Institute	Launched Jan. 2009 to provide R&D core funding and grants for solar PV and CST. Initial budget $100M from Energy Innovation Fund allocation for Jan. 2009-June 2012. $50M supplement for US-Australia collaboration announced Nov. 2010.	Round 1 grants announced in two stages: 11 Dec. 2009 total of $11.1M for 5 projects; 11 May 2010: total of $18.45M for 5 projects. Round 2 grants announced late 2010 amount to about $22M.
Solar Cities	As in Table 4.1 with Moreland (in Melbourne) and Perth added to original five 'cities' by Rudd govt.	As of 1/1/2011, 4.8 MW of PV & 20k of smart meters installed, 9.4k home energy assessments done. Additional funding of $14M over 2 years announced in 2011 budget.
Cities for Climate Protection (CCP)	As in Table 4.1.	Funding terminated by Labor on 30-6-2009.

Table 4.2 continued

Policy/program	Description	Comment
Photovoltaic Rebate Program, renamed under Rudd as Solar Homes & Communities Plan	As in Table 4.1.	Following 2007 election, rebate increased to $8000 per system and renamed. Too successful; means tested from May 2008, then terminated June 2009; replaced with Solar Credits Scheme (see next row) in August 2009.
Solar Credits Scheme	Partial substitute for terminated Solar Homes & Communities Plan. From 10-9-2009, providing increased number of RECs until 2015 for small-scale RE systems.	Funded by electricity consumers, not by government. Solar Credits Multiplier for PV RECs cut twice by Gillard govt in 2010-11 and shortened by 2 years. These cuts could be beneficial, reducing the oversupply of RECs and hence increasing value of RECs.
Solar Flagships Program	Announced in May 2009 budget, funding of $1.5B (of which $1.36B is new) for grants for demonstration of 'up to' four large-scale solar power stations, 2 PV and 2 CST, total 'up to' 1,000 megawatts. Round 1 is for 1 PV & 1 CST. Announcement of successful proposals expected mid-2011. Funding cuts of $220M in 2011 budget.	Funding 'committed' over 6 years from July 2009, including $135M redirected from Renewable Energy Fund and up to $200M from the Education Investment Fund. Round 1 shortlist announced in May 2010: 4 PV and 4 CST proposals. Most of announced funding may never be spent, because of (i) funding cuts in 2011; (ii) negligible market incentives for large-scale solar; (iii) without market incentives, $1 from govt for every $2 from industry & state govt is insufficient for investors; (iv) large size requirement (150-250 MW) for solar power stations almost impossible to finance without market incentives.
Remote Renewable Power Generation Program (RRPGP)	As in Table 4.1.	Too successful; terminated by Labor in all states.
Solar Hot Water Rebate, replaced by Renewable Energy Bonus Scheme	As in Table 4.1.	Rudd government increased rebate to $1600 and then in Jan. 2010 reduced it to $1000; discontinued it 19-2-2010; and then replaced it with new household Renewable Energy Bonus Scheme, still $1000 for solar and $600 for heat pump. In Jan. 2011, $150M of total allocation redirected to flood relief. .
Renewable Energy Development Initiative (REDI)	As in Table 4.1	During both Coalition and Labor reigns, $67M paid in grants to 29 projects. Program closed for new applications in May 2008.
Renewable Energy Equity Fund (REEF)	$18M	Closed for new applications in 2009.
National Solar Schools Program	Grants to primary & secondary schools for renewable energy & energy efficiency.	Cuts of $156M revealed in 2011 budget for 2012-13 & 2013-14. Funding will end 2 years earlier than expected.

Table 4.2 continued

Policy/program	Description	Comment
Australian Centre for Renewable Energy (ACRE), initially named Renewables Australia	Announced in May 2009 budget as an umbrella body to promote development, commercialisation & deployment of RE technologies through a commercial investment approach. ACRE subprograms: REDP; 2nd generation biofuels; geothermal drilling; wind forecasting; electricity storage; venture capital fund; emerging renewables.	Total fund of \$465M = \$100M new + \$365M existing from Howard government's Renewable Energy Development Program (REDP). If REDP takes \$435M, as originally promised by Rudd govt, there would be only \$30M in total left for all the other subprograms.
Advanced Electricity Storage Technologies	Supports development & demonstration of electricity storage technologies for RE.	\$20.4M allocated to 4 projects. Rolled into Renewables Australia/ACRE in May 2009. No new funding announced.
Wind Energy Forecasting Initiative	\$14M. As in Table 4.1.	Provided 'up to' \$14M in grants. Ended 30-6-2009.
Renewable Energy Venture Capital Fund	Launched 10/5/2011. \$100 million ***over 14 years*** to assist early-stage renewable energy companies.	With timespan to 2023-24, it's unclear whether any funding will be available during term of present govt. (But Minister denies long deferral = cut.) Not all new money.
Emerging Renewables Program	\$40M committed in 2010 election + \$60M in May 2011 of unallocated ACRE funds over 5 years.	'Emerging renewables' not defined, possibly geothermal & wave power. It's unclear how much funding will be available during term of present govt.
Renewable Energy Future Fund	Announced in May 2010 budget for development & deployment of renewable energy & energy efficiency; \$652M over 4 yrs.	Details not yet provided by government. It's unclear how much of this is new money and how/if this fund differs from Australian Centre for Renewable Energy and Solar Flagships programs.

Websites (accessed 24/5/2011): Australian Solar Institute <http://www.australiansolarinstitute.com.au>; Solar Flagships Program <http://www.ret.gov.au/energy/clean/cei/sfp/Pages/sfp.aspx>; ACRE <http://www.ret.gov.au/energy/clean/cei/acre/Pages/default.aspx>. **Note:** PV = photovoltaic; CST = concentrated solar thermal; REC = renewable energy certificate.

Table 4.3 State and territory feed-in tariffs

State or territory	Type	Date commenced & period	Max size & annual generation	Rate (c/kWh)	Comment
NSW	Gross	1-1-2010 for 7 yrs	10 kW & 160 MWh	60	In May 2011 NSW govt announced termination of tariffs for new systems and retrospective reduction to 40 c/kWh for existing systems.
Vic	Net	1-11-2009 for 15 yrs	5 kW & 100 MWh; 10–100 kW	60 (credit only); standard retail rate	In July 2010, Victorian govt announced extension of the scheme to limited number of large-scale solar power stations, but change of state govt has left this in doubt.
Qld	Net	1-7-2008 to 2028	10 kW & 100 MWh (1-phase); 30 kW (3-phase)	44	Review after 10 years or 8 MW total
SA	Net	1-7-2008 for 20 yrs	160 MWh	44	
WA	Net	1-8-2010 for 10 yrs	10 kW (1-phase); 30 kW (3-phase)	40	
Tas	Net	TBA	TBA	19.363	Tariff = retail price only
ACT	Gross	1-3-2009 for 20 yrs	10 kW; 10–30 kW	50.05 40.04	Prices for new installations reset annually.
NT	Gross	5-2008	Limited to 225 systems only	45.76	For Alice Springs Solar City project only. Capped at $5/day per system.

Principal source: http://www.aph.gov.au/library/Pubs/ClimateChange/governance/domestic/national/feed.htm. **Note:** kW = kilowatt; MW = megawatt; MWh = megawatt-hour. **Note:** By Oct. 2011 almost these feed-in tariffs had been closed to new entrants.

Recommended policies

No single policy is sufficient to guide the transition to a sustainable energy future. Elsewhere it is argued that Australia needs a mix of pricing, directed funding, regulations and standards, education, information and training, and industry development (Diesendorf 2009, chap. 4; Diesendorf 2011b). Keeping the focus on the energy system, a necessary and sufficient set of specific policies is:

- A ban on all new conventional (that is, greenhouse polluting) coal-fired power stations. Major refurbishments and expansions of existing coal-fired power stations and oil produced from coal and tar sands should be included in that ban.
- A significant carbon price that increases over time. Eventually this will make the proposed ban unnecessary.
- Termination of subsidies to the producers and users of fossil fuels.
- Several additional policies to expand markets for renewable energy until the carbon price is consistently high enough to take over. These policies include gross feed-in tariffs, tax concessions, loan guarantees and a renewable energy certificate scheme like the ERET.
- Regulations and standards for efficient energy use in all buildings and energy-using appliances and equipment.
- Funding for essential infrastructure such as railways and transmission lines.
- Research, development and demonstration (RD&D) of innovative efficient energy use, renewable energy and new transport technologies.
- Increased education and training to provide the skilled workforce for the transition.
- Industry policy to foster local manufacturing of the new technologies.
- Policies to assists low-income households and workers disadvantaged by the transition.

Unfortunately, Australia has only a few of these policies: limited RD&D funding for solar energy; an ERET implemented in 2011; feed-in tariffs for small-scale renewable electricity (only net tariffs in most states); and a small set of minimum energy performance standards for appliances and buildings.

Discussion and conclusion

Renewable energy could make substantial reductions to Australia's greenhouse gas emissions before 2020, given the political will. Indeed, renewable energy and energy efficiency are the *only* genuinely zero-carbon technologies and measures capable of achieving large reductions in emissions from the energy sector on this short timescale. Over a longer timescale, it is now technically possible for

Australia to make a transition to 100 per cent renewable energy by 2050. A recent scenario study even claims that such a transition could be made within a decade (Wright & Hearps 2010), but the very short timescale has been criticised on the grounds that: one of the key technologies recommended, concentrated solar thermal power with thermal storage, is not yet fully commercial; there is a shortage of skilled engineers and other workers; and Australia's manufacturing industry would require rebuilding (Diesendorf 2010). The study makes heroic assumptions in terms of new grid connection of huge capacity between Western Australia and the eastern states, and its treatment of bioenergy back-up to solar power in winter.

Despite public opinion surveys showing that renewable energy has strong public support, it has suffered, under both the Howard Coalition government of 1996-2007 and the Rudd-Gillard Labor government of 2007-2011, from unfulfilled promises, chronic under-funding, promised funds not being spent, termination or reduced funding of a few successful programs, projects designed to look impressive without actually building markets, programs of limited scale, and long delays in implementing promised programs that would have been potentially effective.

The Rudd-Gillard government, in particular, exhibited mastery in the art of creating a confusing multiplicity of programs for renewable energy by: splitting, recombining and renaming them; announcing apparently new programs based partly on previously promised funds; making big promises for the medium-term and long-term future when it may not be in office; and designing fundamentally flawed programs that could not be effective in reaching their stated goals. Examples of the latter are the misnamed Carbon Pollution Reduction Scheme, which is the antithesis of an effective emissions trading scheme, and the version of the Expanded/Extended Renewable Energy Target scheme that passed Parliament in August 2009. Since taking power in June 2010, the new Prime Minister Gillard has either cut or deferred about $400m from the modest funding programs promised by Rudd for renewable energy.

The result is that renewable energy has been tightly constrained in its further development, limited to niche markets and sidelined from its potential as the principal solution to global warming and peak oil. It is difficult to believe that this consistent pattern of blows against renewable energy is due to incompetence alone. A more plausible explanation is that, under the influence of the greenhouse mafia, both the Howard and Rudd-Gillard governments have been deliberately holding back proven renewable energy technologies in the hope that an unproven technological system, carbon capture and storage, could eventually catch up.

Given the failure of both the Howard and Rudd-Gillard governments to implement effective clean energy policies such as those outlined in this chapter, the main hope for building political pressure for a sustainable energy future now rests with the growing climate action movement (Diesendorf 2009).

Chapter 5

Between markets and government: Natural resources management policy in Australia

GEOFF COCKFIELD

Introduction

Natural resources management (NRM) policy in Australia since the early 1980s has been something of an unstable accommodation of two ideologies. Environmentalism drives the demand for the preservation and, desirably, the enhancement of natural capital, while market liberalism drives the demand for limited government intervention and the protection of private property rights. From that intersection, the major national NRM programs developed since 1980 have been largely based on inducing, as opposed to requiring, resource-holders to undertake actions to increase the value of natural capital, although under political pressure there have been some forays into direct government action through regulation and coercion. Four broad strategies are discernible in the suite of programs implemented in the past thirty years: the use of market-based instruments (MBIs) that create a financial incentive for resource protection; the creation of virtual commons, whereby people with an interest in resource protection are encouraged to work together across private property boundaries; the withdrawal of property rights from particular private resource holders; and bioregional governance, whereby an organisation aligned with a region largely defined by biophysical features sets priorities and delivers elements of the major NRM programs. These strategies were variously evident in: the National Soil Conservation Program (1983–1992); the National Landcare Program (1992–2008); the Regional Forests Agreements (1995–2000); the Natural Heritage Trust (1997–2008); National Action Plan on Water Quality and Salinity (2001–08); Caring for our Country (from 2008); the National Water Initiative (from 2004); and the Murray-Darling Basin Plan (from 2010).

There has been, and will be into the foreseeable future, some major expenditure, with: an average of $207 million per year allocated to the NHT from 1996 to 2007 (Australian Government 2007, p. 3; Crowley 2001); projected funding for *Caring for Our Country*, which subsumed the NLP, NHT and NAP, of $310 million per year, based on 2010–11 (Australian Government 2010, p. 6); and a projected expenditure of $1.29 billion per year for the next ten years for a range of water resource protection schemes (DEWHA 2010). There have also been some major changes in the institutional landscape, with: permanent NRM departments or sections of mega-departments in state and territory administrations; a network of catchment management organisations receiving federal government funding of $138 million in 2010–11 (Australian Government 2010); and the Murray-Darling Basin Authority (MDBA) which has, under the *Water Act 2007*, the unenviable responsibility of developing a plan that will increase the level of environmental flows in the Murray-Darling system (MDBA 2010). The ultimate success of this expenditure and these institutional arrangements and new programs would see, with this discussion confined to terrestrial resources: improvements in soil condition; an increase in environmental flows in major river systems and especially the Murray-Darling; and an arrest of the decline in terrestrial biodiversity. In addition, for these national programs to be considered successful there should be an improvement in these three major goals across a broad range of bioregions, and the improvements should be the result of the programs and not other social and economic factors.

There have been improvements in soil management practices, to which some of the NRM programs have contributed, although adoption of such practices is highly variable across farm landscapes. In regard to water flows, the Federal Government has been 'buying back' environmental water, but the major test will be the MDBA's further development of the Murray-Darling Basin Plan and the Federal government's will to implement that plan (see Connell, this volume). Again, though, the proposed policy approach of buying water from willing sellers may result in some differential resource protection outcomes across landscapes. Finally, the greatest failure to date has been the continuing decline in terrestrial biodiversity (ASEC 2006; Fitzsimons *et al.* 2010). The federal government, working reasonably co-operatively with state governments, has laid policy foundations that could achieve much more, even with the present level of funding and working within the federal system, but there are three other limitations. First, the policy frameworks that would lead to the formation of environmental markets and virtual commons have so far been implemented in only weak forms. Second, the form of bioregional governance that has developed in Australia is also very weak, although this does vary a little across the states. Third, state, territory and federal governments seem to be rather timid in taking the lead in completing the conservation mosaic at the inter-regional scale. They are the only bodies that can identify and secure, sometimes with direct acquisition, natural assets such as representative conservation areas, habitats for rare and endangered species, and the bio-links that allow species adaptation (Mansergh, Cheal & Fitzsimons 2008).

The development of NRM policies in Australia

One overarching narrative of natural resources policy in Australia, adapted from structuralist accounts of other western countries (see for example Friedmann & McMichael 1989; Bonnano & Constance 2001; Tilzey 2006), is that of the long dominance of commodity-focused agriculture and forestry eventually breaking down under post-war economic globalisation and the rise of both market liberalism and environmentalism in the 1970s and 1980s. Agri-food policy networks, long the preserve of farm organisations and agricultural ministries, were opened up to various interests wanting to remove trade barriers, reduce subsidies, or ensure that rural landscapes had social and environmental as well as financial values. A slightly more nuanced account, following Wilson (2001) and Evans, Morris & Winter (2002) in their analysis of developments in Europe, is that post-1980 NRM programs are yet another change in the mix of social and economic benefits expected to be derived from the use of natural resources.

In Australia, early land use policy was constructed to achieve a range of social and economic outcomes. The main policy instruments were the assumption, resumption and allocation of property and property rights, and the provision of direct and indirect subsidies. Policy goals at various times included: the reform of emancipated convicts (Connors 1970; Ward 1975); counterbalancing the power of the squattocracy (Roberts 1924); the civilisation of the frontier (Pike 1962; Waterson 1968; Johnston 1988); offsetting the effects of an economic depression (Connors 1970); supplying resources to the industrial hub of the British Empire (Schedvin 1988); and managing social unrest after the gold rushes and the first world war (Callaghan & Millington 1956; Connors 1970; Ward 1975; Lake 1987). The pursuit of some of these social and economic goals, and the instruments of that pursuit, increased the degrading pressures on natural resources above what might have been the case if market forces had been the main determinant of land use. In their efforts to accelerate the development of higher-value agricultural industries, such as the wheat and dairy industries, and to create a class of small-scale producers, state governments, which had direct constitutional control of resources, increased the pressures on those resources in three ways.

First and directly, governments required people receiving land to clear and cultivate it (Roberts 1924; Bromby 1986; Shaw 1990; Lines 1991), thereby increasing land use intensity before the market signals might have done so (Davidson 1981). There still would have been land clearing, given the competition between pastures for grazing and trees (Bolton 1992), but the impacts of cultivation, especially on marginal lands, may have been reduced or at least delayed. Second and also directly, governments built major irrigation infrastructure that allowed for high levels of water extraction from riparian systems. Again, without state support there would have been some private schemes develop over time, but these may have been more dispersed and of a smaller scale. Third, and indirectly, the creation of small-scale farms increased the need for landholders to intensively use natural resources. The parcels of land allocated as part of 'closer settlement schemes' were often too small to be economically viable (Connors 1970), or soon became so (Davidson 1981). The creation of many farm businesses increased downward pressure on prices, so that individual farmers used resources to the maximum. The degradation effects were, at least initially, magnified by ignorance in that the early farming techniques and land use choices were effectively experiments in determining how and where to farm; the consequences, such as soil erosion, fertility decline and soil salination, were not always immediately apparent.

Nonetheless, some of these effects were publicly acknowledged by the late nineteenth and the early twentieth centuries, and there were even some policy responses. The Western Australian (WA) Government introduced a form of soil conservation legislation in 1889 (Bradsen 2000), and in 1917 a Victorian Government committee looked at the issue of wind erosion, although there was no immediate action (Bolton 1992). In 1923 the South Australian (SA) Government introduced the *Sand Drift Act*, under which farmers had some 'obligation' to restrict the effects of wind erosion (Bradsen 2000). Legislative developments in the 1930s followed the policy path taken in the US in the wake of the 'dust bowl' period. The US response, based on honouring strong property rights, was meant to encourage '… individual action through creating awareness and seeking to establish an ethic of stewardship, and through provision of technical and management advice' (Bradsen 2000). This reluctance to curtail property rights was to become evident in later policy responses, but by the 1950s the pressure for government action was somewhat alleviated by improvements in commodity prices and weather conditions so that concerns about soil degradation 'fell off' the policy agenda (Bradsen 2000).

From 1971 there was some renewed concern about soil erosion, and between 1972 and 1984 there were sixteen inquiries relating to the problem, most of which recommended the development of a national land use policy framework (Bradsen 2000), even though oversight of natural resources had previously been a state responsibility. In 1983, the National Soil Conservation Program (NSCP) was launched, whereby the Commonwealth funded state agencies to undertake 'high priority projects' (Lockie & Vanclay 2000). Parallel to this, community groups were forming in order to address a wider range of land protection issues, and the Victorian and WA Governments introduced amendments to formally support and promote such groups, with the Victorian Government's Landcare program establishing the model and the brand more formally. The 1990s was declared, by the Federal Labor government, to be the 'decade of landcare' and a national program was launched in 1990, within the NSCP, followed by The National Landcare Program (NLP) in 1993. In 1997, the Howard Government introduced a more comprehensive natural resources program in The Natural Heritage Trust (NHT), which included the objectives of 'sustainable use of natural resources', 'biodiversity conservation', and 'community capacity building and institutional change' (The Auditor General 2008, p. 34). The NHT covered both private and state-managed landscapes and resources. The scope of NRM programs was further expanded with the National Action Plan on Salinity and Water Quality (NAP), launched in 2000 with the goal of enabling regional communities to address the issues of salinity, water quality and biological diversity (Commonwealth of Australia 2002). The NAP had a much stronger focus on 'catchment management', which has three main features: central policy co-ordination, in this case at the Commonwealth government level; stakeholder participation in program delivery; and planning units based on catchment (watershed) boundaries.

An antecedent to this bioregional management of resources was The River Murray Commission (RMC), which was established in 1917 primarily to manage water allocations (Walker 1994). This remained its primary role for many years. In 1967 the RMC, acting somewhat outside its brief, commissioned a report, eventually published in 1970, on salinity in the river system (Kellow 1992). In 1973, the Federal Government convened a meeting to discuss the River Murray, with special emphasis on salinity (Kellow 1992). From the 1973 meeting there was some agreement on an action plan between four of the governments involved (NSW, Victoria, SA and the Commonwealth), but this was not translated into an agreement until 1981. In 1985 the Murray-Darling Basin Ministerial Council was formed, followed by the Murray-Darling Basin Commission (MDBC), or the advisory body, in 1988. The MDBC was to explicitly consider salinity and drainage beyond the immediate rivers (Walker 1994). The NAP, although primarily focused on hydrological systems and water quality, had a broader scope in considering associated habitat, especially riverine wetlands and the issue of 'over-allocation' of water. With increasingly dry conditions during the early 2000s, demand for urban water and the rise of the Greens in Australian politics, the Federal Government assumed an increasing role in managing major riparian systems, especially the Murray-Darling. In 2004, the National Water Commission (NWC) was formed, with the aim of addressing over-allocation and securing sufficient environmental flows to create 'sustainable' river systems. In 2008, the Murray-Darling Basin Authority (MDBA) took over from the MDBC, and it is required by the Commonwealth *Water Act 2007* to, amongst other things '… protect, restore and provide for the ecological values and ecosystem services of the Basin' (Murray-Darling Basin Authority 2010, p. xii).

While the Murray-Darling system was increasingly being managed as a super bioregion and the Landcare movement and programs had created some local management of shared resources, the NAP and later the NHT formalised a 'middle' layer of NRM governance through the 'regional delivery model' whereby states and territories were expected to have regional NRM organisations and to jointly fund them with the Commonwealth. The idea was that some priorities could best be determined close to the specific problems (The Auditor General 2008). These regional bodies provided extension services, funded collective on-ground works following the Landcare model, and even developed and trialled market-based conservation tenders. NRM tenders have been used in the US since the mid-1980s, after funding formerly used to directly support agricultural production was channelled into conservation programs (Ribaudo & Claassen 2006). The idea is that landholders bid to provide certain landscape outcomes, such as vegetative ground cover, that contribute to the protection of natural resources. Various state government agencies and regional bodies used funds from the NAP's 2002 National Market-based Instruments Pilot Program to run small-scale tenders (see examples in Hajkowicz 2009). In 2007, the Howard Government introduced the first national tender scheme, with the Environmental Stewardship Program (ESP), now part of Caring for Our Country, to focus on particular 'assets' such as box-gum grassy woodlands (Hajkowicz 2009), and with a series of tender rounds intended to work through each asset type. The ESP was partly based on the 2005 Tasmanian Forest Conservation Fund, whereby landholders tendered to provide forested conservation areas.

This Tasmanian scheme was one of several approaches to conserving native forests, and although there has been some recent convergence in the use of such market-based instruments, the development of native forests policies has followed a somewhat different path from those for agricultural resources. The early timber 'industry' was largely based on individuals harvesting on private and public land, and although some landholders directly traded timber (Taylor 1994), the role of private small-scale forestry was and is minimal. State agencies moved to reserve areas for timber production (RAC 1992). They later moved into plantations, becoming major timber producers and processors, between them controlling more than 90 per cent of plantation area by the 1950s, before private ownership of plantations increased from the 1960s (Wood *et al.* 2001). From the late 1980s, Australian state and federal governments started to develop a set of policies to conserve more of the forested landscape (Commonwealth of Australia 1992, p. 3), with increasing timber supply coming from a plantation subsector that would '…develop under normal market conditions as opportunities occur' (RAC 1992, p. 71). To facilitate the transition from native forest to plantation timber sources, the Federal Government took the lead in negotiating Regional Forest Agreements (RFAs) (1995–2000). These twenty-year agreements between the Commonwealth and some of the state governments (NSW, Tasmania, Victoria and WA) were based on designating areas for conservation or for mixed uses that could include logging (DAFF 2008) and included funding for structural adjustment in regions dependent on native forest timber (Herbohn, Harrison & Herbohn 2000). The Queensland Government was not in a formal agreement but also undertook to shift timber production from native forests to plantations.

A key difference between forests and farmland was that state governments retained full ownership of much of the forest lands subject to commercial-scale logging. Much of the pastoral zone is still owned by state governments, but long-term leases and residence on the land seem to have conferred something akin to full property rights on the landholders. Nonetheless, much leasehold and freehold land is now subject to the states' and territories' vegetation management acts, introduced from the 1990s, which define the rights of landholders to clear land so as to maintain regional conservation values (Lockie & Higgins 2007). These legislative measures have been politically difficult for governments, especially the Queensland Government (Lockie & Higgins 2007), just as the withdrawal of logging rights continues to be a political flashpoint. The difference between vegetation and soil conservation as political issues is that forests have long been a staple of environmental activism in Australia. There are votes in preserving forest systems, especially old growth forests. Furthermore, governments get indirect assistance to achieve outcomes from environmental organisations directly targeting major timber companies to the extent that some intend to 'voluntarily' move to an almost complete dependence on plantations (Manning & Darby 2010).

Related to this direct activism, the focus on the environmental impacts of timber companies has led to the development of industry standards. In 2008, under the broad umbrella of the Montreal Process Criteria and Indicators (C&I), Australia developed a Sustainable Forest Management Framework of C&I for assessing the sustainability of forest management through the national-level Montreal Process Implementation Group for Australia. This voluntary certification includes commitments to consider ecosystem health, biological diversity and regional development (AFSL 2007). Voluntary certification has not extended into agricultural industries in the same way. Environmental organisations have, however, increasingly focused on the other resource issues, notably supporting the restrictions on land clearing and, more recently, proposals to significantly cut water diversions for irrigation in the Murray-Darling system (see for example Dunlevy 2010), although these organisations tend to be much more confrontational with major corporations than with farmers and pastoralists.

Policy implementation and outcomes

From this brief review it is possible to identify the four broad approaches to NRM policy in Australia; these are listed below, where they are also considered in relation to degrees and types of government intervention.

1. *Moral suasion and incentives*

At the least interventionist end, state and federal governments have applied some moral suasion and incentives to encourage people to manage virtual commons. Through programs such as Landcare and sub-catchment planning sponsored by NRM organisations, government agencies supply scientific information, targeted funding and, in some cases, facilitators to encourage groups to work as a 'community of resource users' (Schlager & Ostrom 1992). While most land and some water rights in these situations are allocated to individuals or families, systemic interactions mean that there are elements of pooled resources. For example, clearing on one property contributes to salinity on another, or water pollution from one source affects other properties and so on, with some arguing that externalities are best addressed through direct government regulation (see, for example, Ophuls 1973). Ostrom, Walker and Gardner (1992) suggest, however, that groups can be autonomously co-operative where there is a degree of continuous communication and sanctions, which in the case of Landcare projects might comprise social retaliations or the withdrawal of co-operation on other community or neighbourly activities. Hence, government facilitation of local collective action may help to manage resources where there are overlapping interests. Participants may also value the shared purpose and ability to shape local outcomes (Youl, Marriott & Nabben 2006).

On the other hand, Landcare programs have been criticised as being a means for governments to be seen to be acting on environmental concerns while passing the responsibility to landholders (see, for example, Lockie 1993). In addition to this possible tokenism, there are structural and social limitations. There is the problem of group sustainability, in that volunteers, especially in rural regions that are depopulating, are vulnerable to flagging enthusiasm and that there is evidence of generally declining Landcare membership (Youl, Marriott & Nabben 2006, p. 33). In addition, local initiatives do not necessarily contribute to broader strategic goals. The Envirofund was established in 2002 within the NHT. It is specifically for small projects, but there was an expectation that groups would move to a broader engagement with '… regional approaches to NRM' (CIE 2005, p. 19). This has hardly been achieved and it seems that participants did not particularly want or expect that to happen as they focused on particular local concerns or interests (CIE 2005, p. 31). Further to the issue of fragmented outcomes, funding for Landcare-type activities has tended to be broadly distributed, limiting the capacity to concentrate funds on major problems. The Envirofund looks to be impressively funded, with almost $70 million in grants provided from 2002 to 2005, but this was for more than 4,600 projects at an average of $15,000 per project, illustrating the tendency to a 'scattergun' approach (CIE 2005, p. 54). A review of the program found that its achievements in relation to NRM problems were 'quite modest', with some form of rehabilitation, revegetation or protection measures applied to 90,000 hectares (ha) from 2002 to 2005 (CIE, pp. 28–29).

2. *Bioregional governance*

The second broad policy approach has been to develop bioregional governance. It has some of the elements of the virtual commons creation, with efforts to foster identification with a biophysical region in which the benefits of protecting the resources is emphasised. Many of the funded projects and planning exercises also directly support group activities at the sub-catchment level, effectively adding to the Landcare approach. The regional delivery model is, however, also meant to co-ordinate local and regional activities so as to contribute to national priorities (The Auditor General 2008, p. 14) and, as such, this form of governance has a higher degree of central – and particularly Federal – government involvement than is the case for the designated Landcare programs. The regional bodies have a responsibility to achieve particular outcomes, based on their agreed plan, but they have little in the way of direct power, except for some statutory functions for the Victoria and NSW organisations. This, then, is a relatively 'weak' form of bioregionalism compared to stronger models where there is greater autonomy and devolution of authority (Dryzek 1997; Dodge 1999). The regional bodies are another layer of governance amongst the three existing tiers (federal, state and local governments), where little or no authority has devolved to this new layer. Furthermore, these 56 bodies are constitutionally creatures of the state governments and financially beholden to both state and the Federal governments. Since inception, these bodies have generally lived on funding guarantees of no more than three years at a time, so that long-term planning is limited and there is little job security to encourage the development of permanent, regional NRM professionals. Finally, the regional bodies may not have the scientific and evaluation capacity to adjust their strategic plans according to on-ground evidence (The Auditor General 2004, 2008). There has certainly been considerable expenditure, with more than $1billion allocated through the regional bodies, but how much gets to the ground is not clear and there is something of a disconnect between national scientific capacity and its application at the regional level (The Auditor General 2008). It might be argued that a stronger form of bioregional governance is evident in the operation of the MDBA, but on the other hand it could be another extension of centralising federalism. The *Water Act 2007* and the Basin Plan, with MDBA oversight, are to some extent means of keeping state agencies and governments tied to outcomes. There is, here, a dilemma for the Federal Government in balancing demands for national and cross-regional co-ordination against the need for regional adaptation of general principles. It is not clear that the current system of bioregional governance has provided a good balance.

3. *Market-based instruments*

The third broad approach is that of using MBIs to induce change. These can be considered as a set of instruments that range from those with few market principles and high government involvement to those with strong market principles and little intervention. At the higher-government-involvement end of the scale, there are programs where a government is the only buyer, or at least the main funding body, for a limited range of buyers, as with the acquisition of land for the National Reserve System (NRS). The Federal Government funds state governments and other organisations to buy land in 'under-represented' bioregions as it becomes available (Australian Government 2010, pp. 24–31). It is a market-like situation in that sellers are voluntary but this has added to the stock of land permanently reserved for conservation purposes, which now stands at more than 90 million ha, or almost twelve per cent of the terrestrial land area of the country. As part of that, the Federal government is, through voluntary agreements, negotiating with indigenous peoples to create Indigenous Protected Areas within the NRS. Effectively, the Government is purchasing some existing property rights that might, for example, allow grazing, and changing the tenure of that land to conservation uses. More than twenty million hectares, four per cent of the terrestrial land area, are reserved in this way, and the Government is aiming for more (Australian Government 2010). There are three qualifications to what might seem an impressive performance in protecting so much land. It is largely a voluntary program, so it could take some time to fill in the habitat and bio-link gaps across landscapes. Furthermore, there might also be skewed representation of bioregions because the weak (first) sellers will be those with a low opportunity cost, such as the low rainfall pastoralists. Finally, reserved areas need to be managed to achieve the ultimate outcome, which in this case is stopping biodiversity decline, and both fires and invasive pests remain major threats to even protected areas (Fitzsimons *et al.* 2010).

Closer to a recognisable market are environmental tenders and water buy-back schemes. Governments are the main buyers but there is greater competition amongst potential sellers. Advantages of the conservation tenders are that they can lead to the development of conservation management skills in landholders, and can even encourage conservation work above that required by the agreement (Stoneham *et al.* 2003). Like the NRS, environmental tenders can be used to target high value assets, as with the ESP and the Tasmanian Forest Conservation Fund. The weaknesses include the potentially fragmented outcomes with low rates of connectivity, although this latter characteristic can be included as a criterion and heavily weighted. Further to that, market theory would suggest that, as governments try to fill in the landscape gaps, landholders will be able to hold out for much higher prices. These schemes are also highly dependent on the terms on offer, with most schemes to date offering contracts from ten to 25 years, so as contracts expire the previous gains may be eroded. To date, most of the tenders run by state government agencies or regional bodies have been limited in terms of funding, time of the agreement, and area secured (Hajkowicz 2009). At the national level, for example, the Forest Conservation Fund is intended to secure up to 45,600 ha, while the ESP box-gum round may secure up to 27,000 ha.

Market-based instruments have had a much more significant effect on water management, especially the expansion of water trading, but water trading in itself does not necessarily deliver environmental flows (NWC 2010). There may be indirect effects where trading results in higher water prices and the consequent increased water use efficiency increases environmental flows. Alternatively, water might be traded away from high-impact uses, such as cotton production, to lower-impact uses such as domestic consumption. Finally, the higher price might accelerate industry adjustment so that marginal land goes out of irrigated production. The most direct environmental gains come from capping water diversions, which will be done through buying back environmental flows. For the Murray-Darling system, the Federal Government will be the main, if not the sole, buyer on behalf of the environment and will buy from willing sellers, although presumably it will be a selective buyer with the proposed regional targets (MDBA 2010, p. xxv) in mind. The nature of the river systems means that voluntary purchasing at any point can contribute to the health of other parts of the river system, but there could still be some degree of differential regional benefits and it may take some time to get sufficient sellers to reach the Basin Plan target, assuming that there is the political will to do so. The MDBA has proposed the acquisition of an additional 3,000–7,600 gigalitres (GL) per year for the environment (MDBA 2010, p. xix), which would be 20–49 per cent of current diversions. After approximately two years of purchasing, the Federal Government had by the end of 2009 acquired 766 GL (DEWHA 2010), but this may have been picking up the weak sellers in a drought period.

4. *Withdrawal of property rights*

The final approach that state and federal governments have used is to directly withdraw property rights, and these, along with the various forms of reserving, have resulted in some of the greatest changes in resource use. The RFAs have seen the area of native forests available for timber production decrease from 13.4 to 9.4 million ha from 1998 to 2008 (DAFF 2008). The problem here, as with the reserved areas, is that the quality of the ultimate environmental outcomes, such as biodiversity, now depends to a large extent on the quality of the conservation management of these effectively reserved areas. In regard to vegetation management legislation, the net annual loss of forest cover decreased from around 500,000 ha a year in the 1980s, to around 100,000 ha by 2004, with clearing rates generally decreasing and regrowth rates increasing steadily (ASEC 2006, p. 71). Most of these vegetation protection laws now also have a quasi-market mechanism in allowing clearing, where offsetting conservation outcomes are provided in other areas (Burgin 2008). Again, there is a question about the quality of post-protection management, with the additional impediment that landholders were alienated by some of the accompanying consultative and legislative processes (Lockie & Higgins 2007) and so may be resistant to engagement with conservation management on a voluntary basis.

The Labor Federal Government, and perhaps surprisingly the Howard government also, showed preparedness to enact one-off property rights retractions in times of crisis. In 2007, *ad hoc* restrictions on water allocations in the Murray-Darling system were applied to immediately and drastically limit diversions during the millennium drought (1999–2009), with up to a 70 per cent reduction in the area of major irrigated crops (ABARE 2010). Some of this was due to the sheer lack of available water, but some was the direct overriding of property rights to water use. This also illustrates the necessity that the supposed voluntarism of water buy-backs is actually underpinned by regulation. There are caps on diversions from river systems, enforced by penalties for non-compliance, and although the Federal Government has rejected reducing allocations through a decreasing cap, effectively whittling away maximum allocations, the purchase of environmental flows will effectively lower the legally enforced cap.

NRM policy into the future

Australian governments have, since 1980, used an array of policy instruments and institutions to address NRM issues. The argument here is not that these were inappropriate or insufficient, but rather that they could be strengthened and used in much greater concert with each other. This will be particularly important considering that the policy environment might be less favourable in some ways than has been the case in the last fifteen years. Governments have to some extent benefited from some diminishing pressures on natural resources as a result of policy, social and structural changes. First, the reduction in assistance to agriculture, along with international competition and national productivity gains, accelerated structural adjustment processes so that intensive agriculture ceased on some marginal lands. Second, the general economic pressures, combined with findings from state research and extension programs, have seen farmers reduce cultivation and thereby reduce some soil erosion problems through the use of reduced tillage. Third, governments ceased to develop major irrigation schemes, thereby limiting the increase in diversions. Despite a flurry of interest in 'northern development' in 2007–09 (see for example Shanahan 2007; Huxley 2007), the Northern Australia Land and Water Taskforce formed to investigate the possibilities was ultimately very cautious in regard to expanding irrigated agriculture in those regions (NALWT 2009).

Finally, some agricultural land has been converted to less intensive uses as a consequence of the so-called tree-change trend, whereby people buy property partly or solely for lifestyle reasons, usually in the higher rainfall areas and/or closer to major urban areas. Barr (2008) has identified the emergence of transition landscapes, passing from primarily agricultural to more multi-functional uses. This land use transformation is aided by the historical legacy of closer settlement, in that the creation of many separate titles has allowed for easier sub-division. There are some potential problems from an environmental point of view with this trend, including a reduction in land management skills and weed control (Mendham & Curtis 2010), but the reduction in intensive uses reduces pressures on soil and demand for ground and surface water, and in some cases there is some regeneration of vegetation.

Into the future, though, there will be renewed pressures on natural resources. Urban growth continues to be a major form of land disturbance. Then there are the impacts of climate change. Warming and drying may lead to changes in particular habitats, so that important ecosystems might become less viable over time (Dunlop & Brown 2008). There may be an increase in fire frequency and intensity, which could further threaten habitat quality and viability (Hughes 2003). This fire threat will pose a particular dilemma for governments in relation to rural residential areas. Either governments will need to regulate to prevent settlement in fire-prone areas or they will have to relax vegetation control laws. Climate change will also reduce run-off and available water, so that it will be increasingly difficult to maintain environmental flows (CSIRO and Bureau of Meteorology 2007; CSIRO 2008). In addition, the resurgence in mining activity might bring forth new threats, in particular the impact of gas exploration and retrieval on underground and surface water. Finally, if demand for food increases, driving up prices, then there will be an increasing temptation to bring marginal and 'retired' land back into production and to use more water for food production. Hence, there will continue to be pressure on governments to somehow address NRM issues, and there are ways in which the current approaches can be enhanced.

Starting with the least interventionist approach, governments could do more to encourage and strengthen arrangements for the management of virtual commons by combining collective action and environmental tenders. That is, tender schemes could be adapted to allow for vertical and horizontal contracting so that members of a group contract with each other to jointly supply environmental services to a government. This would help to increase landscape connectivity, and would introduce much stronger incentives for peer policing. Then, the Federal Government could go one step further and create much stronger markets for ecosystem services. That is, it could legislate, as happens in other countries, so that users of resources such as water utilities are required to purchase environmental offsets such as riparian vegetative protection (see, for example, Pagiola 2002; Pagiola, Bishop & Landell Mills 2002; Pagiola *et al.* 2004). Further to this, bio-banking schemes could be strengthened to ensure the environmental values of the offsets for the land development are clear and long lasting (Burgin 2008). Given possible transaction costs for 'trading' small areas of land with conservation value, there might be an enhanced role for regional bodies in being environmental brokers, encouraging trades and ensuring accountability. State governments could also further extend bio-banking options within vegetation management laws to allow landholders to negotiate to provide conservation areas and enhancements in exchange for concessions on clearing. This may help improve relations between governments and affected landholders, and could encourage more active land management.

By far the biggest and most immediate challenge for the Federal and four of the mainland state governments is the Murray-Darling water buy-back. This tests both the concept of water markets and the mettle of the governments. Some early indications are that concerns about regional economies and communities, as well as an increased sensitivity to such matters following the 2010 federal election, will result in reduced targets for the environmental flows and careful buy-backs over quite some time. Nonetheless, the mechanism is in place and can be used. Another major policy challenge for the Federal Government could also have some NRM implications. This would be the introduction of an emissions trading scheme, where conservation and plantation forests can be counted as carbon sinks, as was the case with the proposed Carbon Pollution Reduction Scheme (DCC 2009; ABARE 2009). Again, the regional bodies are well placed to become brokers for those with small areas of vegetation that could be carbon sinks.

Moving away from the more competitive MBIs, the Federal government's NRS is removing production-related threats from reasonably large tracts of land. This approach is bedded down and outcomes are a function of willing sellers, income from alternative land uses, and the total available budget. This can operate in conjunction with environmental tender projects, virtual commons activities and the existing system of national parks, as far as filling in the landscape mosaics for environmental purposes. The limitations of the voluntary approach will become more apparent as critical areas and biolinks are identified. The Federal Government will then have to pay above market rates for critical gap areas, compulsorily acquire those assets or, what is more likely, use both compulsion and high payments. The compulsion may need to be limited, though, because management for regional outcomes will require co-operation between land managers operating under a range of different tenures. This leads to perhaps one of the most critical future needs if the ultimate goals, as set out earlier, are to be achieved: these land managers will need the skills and commitment to deliver outcomes for the regions and the nation. That will require training programs and some degree of certainty of career paths. This may be particularly important for remote indigenous communities where labour markets are limited. As part of this concerted effort to improve land management, the most immediate focus should be on combating species loss, which means understanding and controlling fire and invasive species, especially feral cats, as prime suspects for the continuing decline in small mammal numbers (Fitzsimons *et al.* 2010).

Conclusions

Australian governments, under the co-ordination of the Commonwealth Government, have introduced a range of new policy instruments to supplement the earlier reliance on regulation. In a general sense these are appropriate, but with refinement and strengthening they could improve resource protection outcomes. The more market-based approaches have not failed as such, since water has been bought for environmental flows and there is land being managed for conservation purposes under voluntary covenants, including for indigenous people. The problem lies in the under-utilisation of the policy instruments. Outcomes would be enhanced if there were more programs that allowed for group tenders, so that virtual commons are created and supported and maintained through financial incentives. There is some contradiction in having programs strongly based on individual actions, such as most of the environmental tenders, sitting alongside those, such as the Envirofund, that encourage voluntary conservation work, and this is one way of addressing that fundamental contradiction. In addition, non-government land management for conservation purposes would be enhanced if those individual management plans were part of regional and national landscape planning, which is inevitably the function of governments, and especially the Federal Government.

The regional NRM bodies are important links between the Federal Government and local groups and individual landholders, but they need greater security, the freedom to establish priorities for regional issues, and more streamlined reporting and accountability. There is a central role for governments, especially the Federal government, in filling in the landscape gaps by inducements and as a last resort, though not necessarily a long-delayed last resort, by compulsion and compensation. The most coercive actions of recent times—the state and territory land clearing legislation and the capping of water allocations—had immediate effects, which is one of the argued strengths of regulation, but they are blunt instruments which stop some threats to natural resources quality but are not sufficient to ensure that ultimate outcomes are achieved. The eco-authoritarian temptation is to press for widespread and immediate government intervention, but either or both would be costly and highly coercive, alienating the land and water managers of the future.

Governments do not at present seem to have the capacity to manage existing land to the extent that there is an obvious turnaround in the adverse trends in biodiversity. This, though, is still a major policy failure and speaks to the need for urgent action on strengthening the existing policy approaches within a strong plan for future landscapes.

Chapter 6

Flailing about in the Murray-Darling basin

DANIEL CONNELL

Introduction

Should water management institutions be allowed to evolve through short-term defensive reactions in response to evolving issues, or should they be mapped out rationally and comprehensively before development begins? The history of the Murray-Darling Basin (MDB) suggests that the latter is the better option, although the former has often been the norm. Once the incremental path was chosen – by default after the foundation phase of institutional building in the early twentieth century – it then became very difficult to exercise strategic direction over the evolution of water management priorities. In the MDB, the incremental institution-building option has shown itself prone to capture by the stakeholder interests that became dominant as irrigation development intensified. This is significant given that the need for reform has often been in response to problems caused by development activities. The result of the dominance of the developers in the reform process is that there has been continual pressure to protect their interests to the greatest degree possible. Arguably this was a major cause of the failure of the reforms of the 1980s, and it provides the most serious threat to the current reform effort now getting under way with the implementation of the *Commonwealth Water Act* 2007 and the development of the Murray-Darling Basin Plan.

Foundation phase – early twentieth century

It is quite legitimate for stakeholder groups to protect and promote their own interests. Ideally this should happen, however, within an institutional framework that can also take account of the wider interests of society and other issues such as the need to preserve the ecological integrity of the river system so that it can deliver benefits in the medium and long-term future. The initial plan for cross-border decision-making in the MDB – conceived in the late nineteenth to early twentieth century – was based on the Interstate Commission. It would have provided a decision-making framework through which decision-makers at some remove from state political pressures could have made major, difficult decisions from a River Murray basin-wide perspective. Instead, state interests in New South Wales (NSW), Victoria, South Australia and later Queensland have dominated decision-making and been able to insist on minimal coordination and cooperation across state borders. As a result, four separate state-based foci for stakeholder interests became consolidated and able to comprehensively frustrate many of the whole-of-basin orientated reforms attempted in recent decades. Irrigation-development-driven policy and incrementalism are therefore at the heart of environmental policy failure in the Murray-Darling Basin.

It can be argued that the four largely autonomous 'stakeholder interest' zones would not have developed if the original plan for the cross-border management framework for the River Murray had been put in place. The first phase of cross border water management was consolidated in the early twentieth century. According to the legal scholar Sandford Clark, writing in 1971, the River Murray Waters Agreement and the River Murray Commission – the framework and organisation that managed cross-border water issues in the southern MDB until the 1980s – were designed to operate in combination with the Interstate Commission, which was originally meant to be a major part of the Australian federal system. In Clark's opinion:

> *The Interstate Commission was in a position not merely to arbitrate upon the quantitative and qualitative entitlements of the respective states to water in the main stem of the Murray but also to control and prevent more blatant acts of selfishness by the upstream states on tributary rivers. On this assumption it was entirely appropriate that the River Murray Waters Agreement should be confined to the main stem of the Murray where the cooperative financing and construction of storages and locks was to occur, thereby facilitating the regulated delivery of agreed monthly flows to South Australia (Clark, pp. 235-238).*

These plans were frustrated, however, by a High Court decision in 1915 that stripped the Interstate Commission of most of its powers. According to Clark, without the Interstate Commission 'the River Murray Waters Agreement was a totally inadequate vehicle to impose a management regime for the whole Basin' (Clark, p. 237).

1980s – the second attempt to create a strong decision-making system

Despite the failure to put in place a comprehensive system for decision-making in the early twentieth century, cross-border water management in the MDB was relatively successful for a number of decades. Before the locks were constructed and Hume Dam completed in 1931 there were a number of years, such as 1902 and 1914, when River Murray water was too saline to drink. After Hume Dam was built, river salinity was managed successfully until the 1960s. By the early 1970s, however, it had once again become a serious concern. In response, cross-border arrangements were restructured in the 1980s. The new framework was incorporated in a revised MDB Agreement that for the first time included Queensland and the Australian Capital Territory, although not as fully committed signatories. Key elements of what was known as the MDB Initiative were the Murray-Darling Basin Ministerial Council, the Community Advisory Committee to the Ministerial Council, and the Murray-Darling Basin Commission (MDBC). All three were to be supported by the Commission Office. The reforms reflected changing ideas about how public institutions should be organised and operated. There was a widespread feeling that decision-making could no longer be left to small groups of engineers who had spent their careers dealing mainly with water resource infrastructure. Under the new institutional arrangements, the Basin's river system was to be managed to improve biodiversity and sustainability as well as production. The states and Commonwealth governments sent teams of ministers and senior public servants drawn from the agencies that dealt with these often conflicting responsibilities for production and the environment. This brought the environment and agriculture into the institutional fold along with water management.

In principle, given the inclusion of at least two ministers from each of the governments with interests in the MDB on the Ministerial Council, the new arrangements should have been able to make decisions for the MDB as a whole about major issues such as salinity and over-extraction. However, most of the responsibilities incorporated into the new agreement were advisory or discretionary in nature and needed the enthusiastic cooperation of all governments and agencies involved before they could be implemented. This applied particularly to activities outside the River Murray corridor. In addition, the long-established unanimity principle still applied to all decision-making processes, giving the power of veto to any jurisdiction that wanted an item excluded from the agenda or which was dissatisfied with any decision made. Despite these limitations, however, the early years of the MDB Initiative were marked by widespread enthusiasm and considerable achievement.

What went wrong?

On a number of occasions before he retired as Chief Executive of the MDBC in 2002, Don Blackmore publicly reflected on the adequacy of the inter-jurisdictional arrangements in place in the MDB. In 2001, speaking as a contributor to the Australian Broadcasting Corporation's Alfred Deakin lecture series, Blackmore commented that it would not be possible to invariably resolve problems with 'win-win outcomes for all states' (Blackmore 2001). This statement suggested that in the past in order to be successful, as a result of the need to avoid any government applying a veto, proposals had to be designed so that all parties gained something that they could immediately present as a victory to their home constituency. Each negotiation considered in isolation had to result in every participant being able to claim a significant benefit. It has not been sufficient for a jurisdiction to do well on balance, all things considered, in the medium term. This effectively excluded decisions where there might have been overall benefit for the region as a whole, but at some short-term cost to one jurisdiction. Similarly, if the final outcome of negotiations had to benefit all jurisdictions every time, it also excluded the righting of wrongs where one jurisdiction had previously been badly treated compared to another.

In 2004, the late Peter Cullen - a long time member of the MDB Community Advisory Committee, the Wentworth Group of Concerned Scientists, and one of the foundation commissioners of the National Water Commission - reflected on the fate of the 1980s MDB reforms in an interview for the National Library. He said:

> *In the early years after the reforms of the mid 1980s the Ministerial Council was persuaded to commission a series of major investigations whose results created strong pressure for change. Although this approach caused political pain in the states they reluctantly went along with it because of the growing public demand for action to reverse continuing degradation. It was a very good strategy. The states were uncomfortable with it but were wearing it. It could have been made to work if the Commonwealth Government had given its support. Instead it took the view that the Commission (to a large extent meaning the Commission Office) was usurping its role and it was not prepared to allow that to happen. Commonwealth ministers and commissioners felt that it was they who should coordinate and lead change in the Basin. So the Federal Government undermined the Commission by channelling the new money for natural resources through the Natural Heritage Trust and the National Action Plan for Salinity and Water Quality direct to the state agencies, rather than through the multi-government process provided by the Commission.*
>
> *This set up an unholy alliance between the Commonwealth and the state governments in which both for their own reasons were happy to dis-empower the Commission. As a result you had billions of dollars going to natural resources but not through a coordinating body. This was unfortunate because when the money went through the Commission you had all the governments over-viewing each other's investments and they were able to act as a quality control on each other. Under the new arrangement the Federal Government negotiates one by one with each of the states and it lacks the knowledge and experience to effectively scrutinise what each state proposes. As a result the states have been able to take advantage of the situation and run their own agendas. We now have much less scrutiny on the spending than we had when all the governments were sitting around the table. We got so close to getting it right in the MDB but that power relationship undermined it (Cullen 2004).*

As will be described below, through the *Water Act* 2007, the Commonwealth Government has finally been able to gain control of the development of policy in the MDB. Ominously - given Cullen's analysis of the reasons why the MDB

Initiative failed – the new arrangements as they are currently proposed will be based on a network of bilateral agreements.

2010 – the third attempt

The new arrangements put in place through the *Water Act* 2007 involve a very substantial shift in power over policy to the Commonwealth Government. The MDBC was replaced by the Murray-Darling Basin Authority (MDBA) which, in addition to the responsibilities for water-sharing between the states and a range of programs such as those dealing with water salinity previously exercised by the Commission, was also tasked to prepare a Basin Plan by 2011. As already explained, the previous MDB framework had dealt only with a limited range of issues agreed upon through a voting process requiring unanimity. The Basin Plan is to be comprehensive and take account of all issues that threaten environmental conditions and resource security using a catchment-wide perspective that ignores state borders. Perhaps in tension with that aim, however, it will be implemented by ten-year sub-plans that will be developed by each of the states in the MDB. They in turn will shape the various regional and sub-catchment plans within their areas of jurisdiction. Backing the Basin Plan will be substantial payments for compliance from the Commonwealth.

Through the Water for the Future funding package provided to support implementation of the *Water Act* 2007, nearly $6 billion was allocated to support the upgrading of water distribution and irrigation infrastructure in the MDB to equip the region to respond to the challenges predicted to result from climate change. This has been criticised for a number of reasons, not least that the existing distribution of infrastructure to support irrigation (which uses 95 per cent of water extracted in the MDB) reflects expansion that occurred in the wet decades leading up to the 1990s. Understandably, the communities that grew up around this pattern of water distribution would all like to upgrade infrastructure so that they can stay at their current levels of development. But predictions for the future suggest much drier and more variable conditions. However, if the process of adjustment were done with a strong focus on minimising costs and maximising benefits (and not allowing the process to be hijacked by powerful groups protecting vested interests), this would probably mean different agricultural activities would exist in fewer places than now. The battle between these alternate visions of the future is becoming intense, with many communities seeing this as a fight for their very existence.

In addition, in the Water for the Future package just over $3 billion has been allocated to buy back water to be used to improve environmental conditions. In many regions this aspect of the program is strongly contested, even though water is only being bought from 'willing sellers'. Critics claim that the program will cause communities to shrink because the amount of irrigation in their areas will be reduced. There is also concern that a smaller number of irrigators will have to pay for the same delivery infrastructure that previously supplied a larger number of irrigators, resulting in a higher cost per head. At a more fundamental level, there is still widespread opposition – even if it is not often overtly expressed – to shifting water from irrigation to the environment. There is also concern about the impact on water prices of large-scale government purchasing programs.

Central to the new arrangements is the role of the Commonwealth minister designated with the responsibility of approving the Basin Plan, the state sub-plans and many related decisions – albeit subject to advice from the MDB Authority and the new MDB Ministerial Council. What would happen if the key minister were part of a government that was aggressively pro-agricultural development, sceptical about predictions of climate change, and unsympathetic to sustainability and environmental perspectives? That was the situation for much of the time during which the Howard government was in office. Separate from that there is the continued danger that regional or state-based interest groups will still find it easy to undermine whole-of-basin perspectives. The new arrangements are based on the referral of powers from the states, and this can be revoked. But it is not only state governments that will fight for the interests of their state, no matter what a basin-wide perspective might suggest. The Commonwealth Parliament is itself made up of members from the states. When the pressure is on, they can be expected to lobby for state interests just as aggressively as their counterparts in the state governments.

It can be argued that there is nothing in these new arrangements that will make them more resilient than the previous arrangements if the MDBA Basin Plan involves significant political costs to one of the state governments. The *Water Act 2007* explicitly excludes any penalties from being imposed on non-compliant state governments apart from the withholding of funds. That strategy was tried against NSW with mixed success in 2005 when National Competition Policy payments were reduced in response to that state's poor record in implementing water reforms. Consequently, it seems that the provisions for pressing the states to cooperate with a basin-wide perspective are no stronger under the new arrangements than they were under the MDB Initiative in the two decades before the current reforms.

Context for the current water reform program

The MDB is a large and complex region. It is just over a million square kilometres in size. It has a diverse range of landscapes, ecosystems, land uses and climates, and includes over 30,000 wetlands – eleven of which are listed under the Ramsar Convention of Wetlands of International Importance. Divided between the southern and eastern Australian states of NSW, Victoria, South Australia and Queensland, and the Australian Capital Territory – each with their different systems of water entitlements and management – the MDB is home to just under two million people and it supplies much of the water used by another million in South Australia. Those three million people, together with various industrial activities, use about four per cent of the water diverted from the regions' rivers. The other 96 per cent is used by irrigated agriculture. From all sources, the MDB produces approximately 40 per cent of Australia's gross value of agriculture.

Managing the hydrological assets upon which these riches depend is a difficult challenge. As is the case with many rivers that cross state or provincial borders within federal political systems, water policy and management in the MDB is characterised by considerable intergovernmental and interagency conflict, low decision-making transparency, poor accountability, high transaction costs, and *ad hoc* deals that undermine best practice water management. The MDB is highly exposed to the risks attached to what are known as open resources. In 1968, Garrett Hardin published a short paper titled The Tragedy of the Commons in which he argued that it was difficult to restrain over-exploitation of common resources such as shared pastures, fish and water (Hardin 1968). Critics subsequently nominated many examples of successful management of natural resource systems owned in common, and suggested that his thesis was more applicable to open access resources which lack any effective overarching institutional framework able to control and regulate the behaviour of would-be users as a group. In the case of an open access resource, it is in the interests of each individual user to expand their own consumption indefinitely, because any restraint will only increase the volume available for their competitors. The eventual result is the complete destruction of the resource to the disadvantage of everybody. That is the development path currently being travelled by most large international hydrological systems and, until now, that statement has also applied to the MDB.

To counter the threats that face open resources, the National Water Initiative introduced by the Council of Australian Governments (COAG) in 2004 stresses the importance of a whole-of-hydrological systems approach to policy and management. Implementing a whole-of-systems approach that can take comprehensive account of the full range of issues that shape outcomes is easier said than done, however. Since its establishment to champion the introduction of the National Water Initiative in 2004, the National Water Commission has invested heavily in programs to improve water planning, but they have not been able to change the geographical boundaries or institutional divisions that continue to create a splintered planning environment that discourages a whole-of-hydrological system approach. Below the stratospheric layer of statements of general commitment to holistic outcomes, the daily reality of water management in detail continues substantially unchanged.

Some indication of the challenges facing a comprehensive systems approach is provided by a list of characteristics typical of environmental policy issues such as water management, compiled by the policy analyst Stephen Dovers. He argues that they make environmental sustainability problems fundamentally different from other policy issues (Dovers 1997). They occur over much longer time scales and often cut across established administrative boundaries. Poorly defined but finite limits are common, but it is difficult to take them into account within economic systems committed to reducing restraints on growth in the short term. Environmental systems are frequently subject to thresholds that result in significant loss when they occur, but which are hard to predict and difficult to reverse. There is great uncertainty about the likely effects of policy choices when the benefits can be very long term. Many impacts are cumulative and interact with each other, and threshold changes can cause long-established patterns of management to suddenly produce very different results compared with the past. Even more problematic as the level of development pressure grows, it is difficult to take account of ethical and moral considerations created by the conflicting demands of different sections of society. In addition, the sheer novelty of sustainability problems makes them difficult to handle within traditional modes of management.

The policy and management environment in the MDB is complex. It involves complicated ongoing negotiations between a large number of individuals, groups, organisations and institutions, including governments. In practice, decisions are not made through a top-down process, but are the product of cycles of interaction in which the participants have varying degrees of influence but no single voice is dominant. There are two levels of government each with considerable independence from each other. Within governments there are a number of departments that deal with different issues related to water management. Much of the discussion about inter-jurisdictional water management in the MDB gives the misleading impression that interaction between the Commonwealth and state levels is highly structured, but the reality is much more elusive. The independent centres of power provided the Commonwealth and state jurisdictions create focal points around which contending interest groups arrange themselves, moving from one to the other as their members make strategic decisions about alliances and about how best to promote their goals or to block those of others. The Commonwealth Government supplies the bulk of project funds to a variety of recipients, but usually has to rely on indirect processes of accountability to influence implementation.

States have substantial direct regulatory power but limited funds. Regional catchment authorities and water distribution agencies, such as Goulburn Murray Water and Murray Irrigation, are formally subordinate to state governments but have independent corporate standing, often get Commonwealth funding, and have ready access to state and federal parliamentarians representing their areas. Research bodies and research and development corporations can provide findings that can bolster some positions in public controversies, can discredit others, and can sometimes shift the basic assumptions upon which such debates are conducted. Even more politically active are industry bodies and large companies; these organisations are emerging as important political forces as irrigation-based agriculture becomes more business-orientated. There are also non-government organisations, such as the Australian Farmers Federation and the Australian Conservation Foundation, that appeal to the wider electorate and whose support is needed by governments for major initiatives. Banks and commercial organisations such as Woolworths and Coles can also exert significant influence through their lending and purchasing policies. In addition, local governments have planning powers that can play a decisive role at the district level. Largely excluded from all these interactions are members of the general community. They tend to be involved only very intermittently, but when activated *en mass* can be a decisive and unpredictable political force.

For political parties, contemporary water management creates divisions that ignore traditional party lines. There is a low capacity to organise the economic system to penalise those who are responsible for negative environmental impacts, or to provide rewards to those who incur extra costs by adopting more sustainable management practices. Despite the growing interest in the so-called triple bottom line approach, accounting systems are ineffective in capturing the environmental/social costs and benefits involved. Further, the legal system is frequently unable to penalise those people responsible for negative environmental impacts and the potential benefactors of remedial action often cannot be identified clearly enough to target the collection of costs. Robust criteria for determining whether particular water policy issues should be treated as public or private are hard to develop. A significant underlying factor fuelling the growing crisis is that the predictive capacity of traditional science continues to be limited. Trends and general biophysical processes can be documented at the larger scale, but it is often difficult to identify the links between particular actions and specific consequences with enough certainty to give managers, the courts and the people who will be affected by their decisions confidence in the process or in the results.

The role of water markets

Markets have played a major role in the transformation of the MDB over the past decade, but it is not a story of unmixed success. In 1994, COAG approved a new rural water reform package that was the precursor to the NWI ten years later. Hydrological systems were to be made sustainable, financially self supporting and economically more productive. Markets were seen as the source of energy that would drive water reform. The vision was that a more robust approach to pricing policy would see all or most of the costs of water management - including, somehow, the cost of dealing with environmental degradation - included in the price. Then, the market would move water away from zones of high impact where environmental regulation limited its uses, to areas where it could be applied with less restriction to support higher-value activities, whether they be different types of irrigation, mining or urban centres. This would reduce environmental damage from waterlogging and rising saline groundwater tables - a major policy concern in the 1990s - and increase the financial returns to society as a whole from water-based activities.

There have been some triumphs. Annual reports of the Australian Bureau of Agricultural and Resource Economics on the productivity of irrigation-based agriculture in the MDB have shown that the drought caused little or no reduction to economic returns in the MDB. This is because the impact of the drought has been ameliorated by the way markets allowed water to move away from low-value to high-value activities. From a broader perspective, however, the picture is more complicated. First, it has been difficult to include many important costs in the water price. Second is the problem of stranded assets and of the economic impact on people who were dependent in different ways on the economic activity but who were not able to benefit from the sale of water entitlements. (Good statistics are rare in this area, and much of what is said is hard to prove.) Support for markets is also highly selective. People often approve of their use in some situations and not others. Many people in the MDB support trading between irrigators but not to urban centres (particularly outside the MDB such as Melbourne).

Public participation

These conflicts show the importance of effective processes to promote public participation and decision-making that takes account of community values to prevent conflicts from spiralling into highly confrontational situations that can paralyse water policy and management. Analysis of what is required for effective stakeholder participation in water management has been undertaken most thoroughly in the context of the European Union's Water Framework Directive (WFD). In terms of both scale and ambition, this is probably the world's most ambitious water reform program. The goal is that all major hydrological systems in the European Union (EU) should be of high ecological status by 2015. The unit for policy and management is the river basin. With its emphasis on the need to manage across both political and institutional boundaries, it has many similarities with Australia's National Water Initiative, in particular its whole-of-hydrological system approach. The WFD gives public participation and the role of civil society a central place in its approach to implementation. In broad terms, there are three ways to implement change: by force and negotiation under coercion at the direction of governments; through trade (the market); or through deliberation, education and collaboration. The Europeans are using all three ways, but they place a very strong emphasis on the potential of the third category known under the catch-all term 'social learning' or 'sustainability learning'.

The social learning approach of the EU aims to promote cooperative relationships and behaviour and new attitudes to the environment. In terms of the aims of its proponents, it is not just about solving particular technical problems and achieving specific policy goals. The more general aim is culture change. The importance of public education about water and the environment is a major theme. People are encouraged to meet often in constructive circumstances in order to develop positive relationships, which can then influence the way in which they work on water management problems together. Governments are encouraged to involve stakeholders early in the planning process and provide substantial support to their joint activities. The list of objectives is impressive (although there is no doubt that, even in the best examples, reality usually falls far short of the aim).

As part of its effort to promote public participation, the European Union commissioned the HarmoniCOP project, completed in 2008, to provide member countries with detailed guidelines backed by extensive research (Tabara 2005, p10). The project produced a number of major reports. The Integration report states that four issues were central. These were:

- the need for structural change to water management institutions to create positive conditions for social learning;
- the importance of cultural factors shaping behaviour and the understanding of interactions between people, water and the larger environment;
- the central role of education and effective communications; and
- the need to see social learning as a key element of a wide range of processes involved in learning how to live and work sustainably.

The EU program also stressed the importance of context, and the need to build on established processes and existing networks. Working within existing systems increases the potential to integrate the new with the old, and increases the likelihood that the reforms will be accepted as legitimate. At the same time, the program stressed the need to document what was being done so that different regions could learn from each others' experiences. It emphasised that social learning for river basin management should not be viewed in isolation: instead it should be seen as part of a much wider social movement. At the same time, the report warns that many of the constraints on social or sustainability learning are structural and deeply cultural, and that the change will need to be much more profound than can be achieved directly through governments using command and control methods to implement the WFD or any other legislation. The research by the EU into implementation issues is of particular interest because of the great variety of governance environments within which the WFD is being applied. This caused its planners to go back to basics and general principles, which are likely to be of much wider relevance than just to the European Union. Given that, it is notable that that there has been much less emphasis in Australia on the significance of culture change, education and public participation for water reform. Compared with Europe, water reform in Australia has been treated largely as a technical managerial exercise rather than as a cultural project.

Lateral options

There have been eight major attempts in the MDB since the late 1980s to substantially reduce the volume of water extracted for irrigation and to provide more for the environment. These are the Salinity and Drainage Strategy of 1989, the Natural Resources Management Strategy of 1990, the Cap of the mid 1990s, the 1994 COAG rural water reform package, the Integrated Catchment Management Policy Statement of 2000, the Living Murray First Step project of 2003/4, the National Water Initiative in 2004, and now the *Water Act* 2007 with its Basin Plan which is to be finalised in 2011. All of the previous seven depended on brittle intergovernmental agreements bound together by weak arrangements for compliance. Despite assurances to the contrary, this is also largely true of the Basin Plan. However, there is one innovation contained in the *Water Act* 2007 that involves a different approach, and that is the creation of the Commonwealth Environmental Water Holder (the CEWH) which is to manage the water being purchased from willing sellers by the national Commonwealth Government.

In the coming years, it is likely that the CEWH will emerge as the most important water management institution in the Murray-Darling Basin. The Labor government has made a commitment that all water reclaimed through the Basin Plan will be purchased at market prices from willing sellers, or will come from investment in infrastructure improvements. This water will be lodged with the CEWH to be managed as entitlements for the environment. This means that the timetable for developing the state sub-plans will be almost irrelevant. In reality, the program will be driven by the CEWH and only limited by the size of its budget and any limitations that may be imposed by the Commonwealth. When the CEWH's purchasing program is complete, it will hold more than a quarter of all water entitlements in the MDB. It will be used unfettered by constraints beyond the requirements of the Environmental Watering Plan to be developed by the Murray-Darling Basin Authority, another Commonwealth agency (albeit in consultation with the states).

The significance of the CEWH has been greatly enhanced by the resolution of an apparently obscure debate, about whether the environment was best served by 'rules-based water' or entitlement water, that was conducted at the same time as the *Water Act* 2007 was being developed. The standard way to manage rivers in the MDB has been to develop plans that allocate a proportion of the flow to entitlements holders, with the balance left in the river. Water left in the river is called rules-based because it is the result of applying the management rules. But over the years it has proved difficult to protect rules-based water from encroachment from entitlement holders. The rules have always favoured entitlements when water supplies are reduced during droughts. The management rules have this bias because the original reason for building dams and regulating rivers was to supply water to agriculture and towns during the summer and in droughts when the natural flow is low. In addition, there is the burden of water theft, which Peter Cullen guesstimated was responsible for at least twenty per cent of all extractions. Most of that water comes from the rules-based component of flow.

When they were first introduced in the early twentieth century, water allocations were very loosely defined. In recent years they have been tightened, sharpened and designated as 'entitlements' and 'rights'. As entitlement holders increased their certainty and reliability, the risks of non-supply in over-allocated systems in a highly variable climate became increasingly concentrated within the proportion of flows supposedly reserved for the environment – the rules-based component. In response, some environmental advocates argued for buying water entitlements in the same way as other stakeholders such as irrigators. That way, they would have water for the environment with legal and security characteristics that are identical to those of other entitlements. But in the absence of substantial funds, these scheming dreams were little more than fantasies.

The situation was ironically transformed by conservative Prime Minister John Howard's $10 billion water reform package, announced in January 2007, which included $3 billion for the purchase of environmental water. Although it looked like a radical green gesture to some people, it provided a way to reimburse irrigators for water that they were going to lose *without* compensation if the NWI was implemented in its original form (NWI paragraphs 48-49). Under the traditional approach to water management, where the environment is supplied through rules-based water, most of the water purchased from irrigators would have been returned to the entitlement pool during times of drought because of the way the management rules work. To those who were sceptics of the campaigns to restore the MDB, this might have looked like a win-win situation. The irrigation community would get $3 billion in compensation, and then get most of the water back anyway in times of drought.

The shift to supplying the environment through the use of entitlements changed that dynamic, however. Under this arrangement, water for the environment will be permanently outside the consumptive pool. It will continue to be available to the environment even in times of severe drought (unless there is political interference with the operation of the CEWH). The CEWH and its budget are now controlled by the Commonwealth Government, which will therefore be able to achieve the environmental flow targets of the Basin Plan even if the states do not give their support.

Conclusions

The situation in Australia's Murray-Darling Basin is likely to be of interest for water reformers in other federal political systems. Many of the most influential manifestos for water management reform - such as integrated catchment management, integrated water resource management and Agenda 21 - have stressed the need for ever-increasing degrees of coordination between policy-makers and stakeholders. That has been the case with the many attempts made in the MDB during the past few decades, and is also true of the Water Act's proposed Basin Plan. Almost by accident, however, through the creation of the CEWH, the Water Act contains a plan B, which has the potential to promote a very different pattern of operation. Given the limited success with the implementation of reforms attempting to integrate management systems, despite many attempts in Australia and elsewhere, it is worth asking whether more can be done by creating institutions less dependent on cooperation between all governments. It may be that a river rehabilitation program based on an institution such as the CEWH would be more in keeping with the realities that characterise complex pluralist political systems, with their competing jurisdictions and many stakeholders, such as exist in Australia and in many other nations. In practice the requirement for coordination, cooperation and unanimity empowers opponents of change. Under the arrangements that have been tried repeatedly in the MDB, such opponents have been continually able to support reform in principle but undermine it practice. Arguably, it is now time to develop institutional systems that will allow reformers to implement changes despite opposition. Most areas of public policy are controlled through institutions that allow decisions to be made and implemented even though there is often strong minority opposition. It is time that water policy was subjected to the same discipline.

Chapter 7

Disintegration or disinterest? Coastal and marine policy in Australia

GEOFF WESCOTT

Introduction

Australia is a coastal maritime nation. It has almost 70,000 kilometres of coastline, one of the largest Exclusive Economic Zones (EEZs) in the world, and 85 per cent of its people living in coastal towns and cities. The nation's economy and culture is dominated by the coastal and marine environments. It might, then, be expected that the environmental policies concerned with planning and managing this vital coastal/marine region would be well formed, sophisticated and clear.

A further expectation might be that Australia would be an exemplar in the field of integrated oceans and coastal management – the dominant global concept for guiding marine and coastal policy (Cicin-Sain & Knecht 1998) – and that adherence to this concept could hence be used as a measure of the effectiveness of marine and coastal policy in Australia. Integrated ocean and coastal management is based on integrating, rather than simply coordinating, policy, planning and management across the coastal/marine interface. This area is often called the coastal zone when dealing with coastal land and coastal (near-shore) waters in recognition that no rigid land/sea boundary exists in the natural biophysical environment; rather, there is a transition zone between purely marine and purely terrestrial environments.

In fact, in the more refined and highly developed global concept of integrated coastal zone management (ICZM) Australia has been seen in the past as a world leader (Sorenson 1997). ICZM is internationally usually referred to as Integrated Coastal Management (ICM), but in the Australian context this can be confused with the more common use of ICM – Integrated Catchment Management. As Harvey and Caton (2003) quote from the Intergovernmental Panel on Climate Change (IPCC): 'Integrated coastal zone management involves the comprehensive assessment, setting of objectives, planning and management of coastal systems and resources, taking into account traditional, cultural and historical perspectives and conflicting interests and uses; it is a continuous and evolutionary process for achieving sustainable development' (1994, p.40).

More recently ICZM has been recognised as having five key operating principles (Cicin-Sain & Belifiore 2005):

- intergovernmental (vertical) integration – between the tiers of government: federal, state and local in Australia;
- inter-sectoral (horizontal) integration – between the various sectors operating at the same level of government, for example between fisheries, conservation and oil exploration and extraction;
- spatial integration – between catchment, coasts and marine environments;
- science–management integration – management responses and directions should be based on the best available scientific information at the time; and
- international integration – co-operation and integration between and across national boundaries and systems.

Hence 'policy success' in the coastal and marine domain could be defined as consistency with internationally recognised ICZM, by implementation of the five contributing principles above.

Yet, as the second decade of the twenty-first century begins, despite four major national coastal inquiries since 1980 (Harvey & Caton 2003), the most recent reporting in 2009, the Australian Government lacks a coastal policy or any integrating coastal governance bodies, operating only a loose 'framework' developed in 2006 (NRMMC 2006). The same government possesses a dormant Australian Oceans Policy (Commonwealth of Australia, 1998 a and b), and has split its marine and coastal policy areas into separate government departments.

In effect, current national arrangements are not integrated either horizontally (that is, between sectors/agencies or across the catchment–coastal–ocean environments) or vertically (integration between federal, state and local government).

How has this environmental policy failure developed? What is the immediate future for marine and coastal policy in Australia? Are there any initiatives or positive signs for enhancing this critically important policy domain? The approach adopted below is to briefly describe the origins and background to the current policy status of these two areas (coastal and marine), to analyse the failures in these areas, and then propose a way forward to address these policy failures.

This chapter can be only a brief overview of marine and coastal policy and should not be regarded as a comprehensive review of all policy failings in the marine and coastal domains, for example, in fisheries policy, protected area policy, oil and gas exploration and resource allocation policies. These cannot be included due to space limitations.

Background to current coastal and marine policy in Australia

Under the Australian Constitution and the later Offshore Constitutional Settlement agreement (1979) between federal, state and territory governments, Australian coastal land and Australian coastal waters out to three nautical miles (approximately five kilometres) offshore (the coastal zone) fall under the jurisdiction of state and territory governments. This is consistent with the overall constitutional position in Australia, where land planning and management of nature and natural resources is a state/territory rather than a federal responsibility.

Yet, under the United Nations Convention on the Law of the Sea 1973-1982, Australia now also claims a marine territory (or Exclusive Economic Zone) that goes out to at least 200 nautical miles offshore (not only from the main continent but also from offshore islands such as the Heard McDonald Islands near Antarctica and Macquarie Island to the far south of Tasmania). The resulting Australian EEZ is twice the area of the Australian land mass – one of the largest in the world.

Under the UN Convention on the Law of the Sea (signed by Australia in 1994), Australia has first use of the resources in this vast EEZ (although if Australia is not going to use these resources they are to be licensed, up to sustainable limits, to other nations). As well, Australia takes on the role of protecting this massive marine environment, which stretches from deep into the tropics to Antarctic waters (see Haward & Vince 2008 and Wescott 2000 for more details).

The Australian EEZ, or Commonwealth marine waters, is solely a federal government responsibility and falls under the *Environmental Protection and Biodiversity Conservation Act* 1999 (EPBC Act). Australian policy concerning these environmentally continuous coastal and marine domains – one dominated by the states (coastal zone) and the other exclusively a federal government responsibility – has developed in distinctly different ways and at quite different rates. The development of the two areas will be described briefly in order to elucidate some of the ongoing constraints on policy development in these important domains.

Coastal policy in Australia

Australian coastal policy development

Thom and Harvey (2000) identified four key factors that they believed had been influencing Australian coastal reform at the turn of the century:

1. global environmental change;
2. support for sustainable development;

3. adoption of integrated resource management; and
4. greater community-based management and participation.

That paper was written against the background of coastal policy reform, which was occurring predominantly at the state level through the 1990s and early 2000s but has since stalled. In the federal sphere, the Resources Assessment Commissions' (RAC) Coastal Zone Inquiry (RAC 1993) was a benchmark for coastal policy development, but its major recommendations for a more substantial federal government role, and institutional arrangements to ensure this occurred, were never fully adopted.

In the mid 1990s, a Commonwealth (as distinct from national) Coastal Policy resulted from the RAC inquiry, and funding arrangements under the Coasts and Clean Seas Initiative did see significant coastal works occur across the country, particularly through matching state funding arrangements called 'Coastcare'.

Thom and Harvey's four factors provide a means of analysing coastal policy reform, and can be compared to what have been reported as the more recent 'drivers' of current Australian coastal policy by Wescott (2009).

- *Global environmental change* has narrowed over the last decade to focus on the impacts of climate change on humans, and this is quite possibly the primary driver of current federal government action on coastal policy.
- *Sustainable development* (or ecologically sustainable development in Australia) has certainly not maintained its momentum over the past decade. The more obscure concept of sustainability - mainly taken as 'triple bottom line' outcomes – has not played a significant role in coastal reform at the national level.
- *Integrated resource management*: in contrast to sustainable development, integrated coastal zone management (ICZM) both forms the basis of the current major national coastal policy framework (NRMMC 2006) and is specifically mentioned in state strategic planning documents as a core theme, for example in the Victorian Coastal Strategy (VCC 2008).
- *Community participation and input* has been substantially maintained as a driving force in coastal reform, but this may have become more of a political (electoral) concern than a community-based, management-driven demand.

In the early part of the 2000s there was little new national activity on coastal policy (Harvey & Caton 2003). But by mid-decade two major community-led forces were becoming more active: the 'seachange' phenomenon and the potential impacts of climate change on coastal communities (Wescott 2009). Nevertheless, in 2006 a national framework for integrated coastal zone management was approved by the Natural Resources Management Ministerial Council (NRMMC 2006): not quite the national coastal policy promised by the Coalition government before the 2004 election, but nevertheless an acknowledgement that there was still residual community interest in the federal government playing some role in coastal management in Australia.

The 'framework' identified the key issues confronting coastal environments and communities in Australia on which it suggested there should be 'national collaboration': land- and marine-based sources of pollution, managing climate change, introduced pest plants and animals, the allocation and use of coastal resources (taken to include the vexed question of coastal planning), and capacity building. The 'framework', though, did not suggest a mechanism (that is, any institutional arrangements) for the implementation of ICZM to address these issues, nor did it provide any increased funding to stimulate implementation. It was, in effect, 'policy without implementation'.

The community pressure, that had finally led the federal government to at least produce a national 'framework' in 2006, was particularly active in two specific areas: the 'seachange' (population movement) phenomenon, and the potentially devastating effects of climate change on the coast (Wescott 2009).

The first was stimulated by the increasing profile of the National Sea Change Task Force (NSTF) – a coalition of local government councils in essence lobbying for increased federal financial assistance to fund the increasing number of people moving to regional coastal areas, particularly in south-eastern and south-western Australia. The arrival of these people in small coastal communities, when combined with the servicing of increased tourism visitation, was severely straining councils' resources which were largely drawn from their local ratepayers. Federal government financial involvement, in particular, was seen as a solution to the plight of local governments. Not only did the NSTF act as a lobbyist, it also organised forums and stimulated discussion on a range of potential policy solutions; the various arguments presented for various national coastal policy outcomes and the institutional arrangements proposed are comprehensively described in the book edited by Lazarow *et al.* (2006) based on reports commissioned by the task force.

Meanwhile, despite a federal government (Howard Coalition government) that was not convinced about the reality of human induced climate change, a number of non-government organisations (NGOs), such as the Australian Conservation Foundation (ACF), were gaining substantial community support for action on reducing carbon emissions in Australia and the actions required to rapidly adapt to the impacts of climate change. These impacts on the marine and coastal environments include rising sea temperatures, rising sea level, ocean acidification, and increased storm and cyclone intensity and activity. This was a second major factor in stimulating reform.

A third factor that was building momentum for national coastal policy reform was a change in the electoral dynamics in regional Australia. This change was highlighted by the comments of the incoming treasurer after the 2007 federal election victory of the Australian Labor Party, who remarked that Labor had won office on the back of winning the 'sunbelt seats'. These seats are in fact the regional (non-capital city) seats that abut the coastline from Cape York to the Victorian–South Australian border.

Closer analysis of the final voting distribution did highlight (Wescott 2009) that in 2007 Labor gained seven of these 33 seats, so that the Rudd government now held 48 per cent of these seats (the Liberals held 33 per cent, the Nationals five per cent, and there was one independent). More pointedly, after the 2007 election sixteen of these 33 seats were now marginal and hence their electors were of great interest to the major parties. In the 2010 election, 28 seats returned their sitting member (including Labor holding on to three of the seven marginal seats they won in 2007), but with Labor losing four to the Coalition, and the Nationals one to an independent. Clearly, these regional coastal seats are now more dynamic than in the past, and this can only lead to a closer focus by the major parties on regional coastal issues - such as climate change impacts and coastal development.

Therefore it is not surprising, with the benefit of hindsight, that Labor had gone into the 2007 election with quite a comprehensive coastal policy platform. They duly implemented most of the major actions in this platform in their first term, including establishing a House of Representatives inquiry into the impacts of climate change on coastal communities - hence neatly addressing all three of the drivers of coastal policy described above (climate change, coastal development and electoral politics).

In March 2008, the Minister for Environment Protection, Heritage and the Arts (Peter Garrett) and the Minster for Climate Change (Senator Penny Wong) referred to the House of Representatives Standing Committee on Climate Change, Water, Environment and the Arts (HORSCCCWEA) a brief to investigate the potential impacts of climate change on Australia's coastal communities. Whilst the committee's terms of reference focused on climate change impacts, there was also a reference to having regard to 'governance and institutional arrangements for the coastal zone', and hence it had the licence to comment on how the Australian coast should be planned and managed. In effect, this was another federal coastal zone inquiry. The committee received 128 submissions, held 28 public hearings from July 2008 to August 2009, and engaged in nine site inspections. Their final report, tabled in parliament in October 2009, contained 47 specific recommendations and included 350-plus pages of information (HORSCCWEA 2009).

The majority of recommendations were concerned with climate change, but there were also recommendations on coastal databases, emerging issues and insurance proposals. There were ten recommendations concerning coastal governance arrangements, which in essence proposed establishing an intergovernmental agreement on the coastal zone, which would form the basis for a national coastal zone policy and strategy, with a supporting management unit and a national coastal advisory council. They left open for further development the proposals put to them for a national oceans and coast act and a statutory coastal council.

In effect the report recommended a much more significant federal government role in coastal policy, and a clear and detailed national coastal policy/strategy with strong institutional arrangements – although it stopped short of recommending federal coastal legislation.

Current state of Australian coastal policy

In early 2011, Australian national coastal policy is linked by a framework for co-operation as part of a weak agreement between the Commonwealth and states and territories developed as part of the natural resource management processes in the early twenty-first century (NRMMC 2006). As already noted, this is 'policy without implementation', that is, it is the bare bones of key management challenges for coastal policy in Australia linked by the theme of integrated coastal zone management, but lacking the institutional framework for implementation and put forward without any new funds to implement the proposed integration. Given these implementation flaws, it is not surprising that it has progressed little since the policy was published.

In the context of the litany of policy failures in the national coastal domain over many decades, the Federal Government's response to the House of Representatives' report was always going to be crucial to determining the immediate future of coastal policy in Australia. The Federal Government's response in November 2010 was disappointing. The Government will reconsider the 'framework' and work towards an Intergovernmental Agreement on the Coast, but it rejected any proposal for legislation. As well, the Government seems to believe that the pre-existing Coasts and Climate Change Council was adequate to cover coastal matters at the federal level. It appears that it will continue to view coastal matters as an extension of climate change issues, and, with the Council and the coastal bureaucracy inside the Department of Climate Change and Energy Efficiency whilst marine matters are tackled in a different department, the possibility for an integrated marine and coastal policy now seem as remote as at any time in the past decade.

In conclusion, Australian coastal policy is still struggling to reconcile what the Federal Government's role should be in the domain of coastal planning and management almost twenty years after the landmark RAC Coastal Zone Inquiry. The main motivation for any continuing changes will be the Gillard government's political necessity to convincingly respond to climate change in its first electoral term; if ongoing coastal policy is closely tied to the climate change response, it is possible that a clear federal government role in coastal planning and management in Australia could still finally emerge.

Marine policy in Australia

Australian marine policy development

In 1994, Australia ratified the United Nations Law of the Sea Convention, which extended Australia's Exclusive Economic Zone, or Commonwealth marine waters, to at least 200 nautical miles offshore. Prior to this ratification, Australia had planned and managed its offshore resources in a straightforward sector-based approach; that is, the federal fishing agency managed fisheries, oil and gas exploration was handled by another division, marine protected areas (marine parks) were run out of an agency responsible to a different minister, foreign affairs and defence were managed by their own departments, and so on. Hence there was no coordinated marine policy as such, simply a series of agencies with policies covering their specific sector. But the claiming of a large EEZ requires a government to demonstrate that it is managing its newly-acquired areas responsibly. Some nations, for example Canada, have demonstrated this by passing legislation (an Oceans Act) whilst Australia decided to prepare an 'integrated and comprehensive' Oceans Policy.

Preparation of the Ocean Policy began in March 1997 and, after a series of discussion papers and consultations (Wescott 2000), the final document – *Australia's Ocean Policy* (AOP) – was released in December 1998 (Commonwealth of Australia 1998a and b). This was a remarkably quick development for a policy, and highlighted the importance placed on the newly acquired EEZ by the Australian Government – and in particular by the sponsoring minister, Senator Robert Hill. The AOP included a suite of institutional arrangements which included an overriding National Oceans Ministerial Board (NOMB – containing Ministers with marine portfolio responsibility) and a National Oceans Advisory Group (NOAG), where over twenty stakeholder groups were represented (see Wescott and Fitzsimons 2011 for full descriptions). These bodies were supported by a National Oceans Office (NOO) reporting to the Chair of the Board (Senator Robert Hill). NOO was located in Hobart alongside other marine research institutes under a complicated deal which was linked to the independent Senator Harradine voting in favour of the Goods and Service Tax in the Senate in the late 1990s.

The Oceans Policy was to be delivered through existing agencies and through the preparation of Regional Marine Plans (RMPs) covering major offshore regional ecosystems (Haward & Vince 2008). The first plan was commenced in 1999-2000 to cover the south east region and, despite it abutting the coastal waters of four states (Victoria, NSW, South Australia and Tasmania), none of the state governments co-operated with the Commonwealth in developing the plan. None of these arrangements for developing and implementing the Oceans Policy were supported by legislation.

Wescott (2000) pointed out that there was an inherent contradiction in the AOP, in that it was written with an emphasis on integration of the management of the EEZ, but it also stated that the sectors need not be concerned because they would be continuing their normal pre-existing arrangements. In reality, this contradiction was confronted in the preparation of the first Regional Marine Plan (NOO 2004), which took five years to complete – that is, three times longer than had the preparation of the entire AOP. The result of such slow progress was a change in the institutional arrangements in 2005–2006 in which the NOMB was dissolved and the NOO incorporated into a division within the environment department in Canberra. This was subsequently followed by a change in the method of development of the RMPs, with section 176 of the *Environmental*

Protection and Biodiversity Conservation Act 1999 (concerning the production of Bioregional Plans) used to prepare what were now to be Marine Bioregional Plans. This did give the plans themselves a statutory basis. NOAG continued but it was now responsible to a single minister (of environment).

The perception resulting from these changes was that this was a downgrading of the importance of the AOP, and that it had become an environmental policy run out of a division of the environment department rather than a comprehensive and integrated policy covering all sectors responsible to a ministerial board. This was always going to lead to perceptions of (environmental) bias, particularly in the area of Marine Protected Area (MPA) development, although conservation groups had not requested these new arrangements and were overall very disappointed with the MPAs resulting from the South East Regional Marine Plan (see Wescott & Fitzsimons 2011).

Finally, under the Rudd Labor government, the NOAG was officially dissolved in early 2009, having not met since mid-2007. This completed the almost entire dismantling of the institutional arrangements introduced to implement the AOP in 1999-2000. As a footnote, the Marine and Coastal Community Network (MCCN) - an information broker with 10,000 participants established by the Hawke government in 1993, which had played a key role in stimulating community participation in marine policy - had its funding removed by the Labor government in 2008. It also has not been replaced.

Current state of Australian marine policy

The Labor governments of Rudd and Gillard have not introduced any new arrangements to replace the previous ones described above, and have made no suggestion that they will in the near future. This current state of Australian marine policy was highlighted by a bizarre series of events during the 2010 federal election campaign. The opposition leader, whilst in Townsville courting votes in a fish market, accused Labor of running a regional marine planning process that seriously disadvantaged fishers (commercial and recreational) and stated that the Labor government had not consulted user groups. He claimed that there were plans for massive 'no take' marine parks and that the Labor government had 'locked up' large areas in MPAs since being elected. The reality was quite different. Labor, since being elected, had continued the bioregional planning process designed and introduced by the Howard Coalition government under legislation developed by and implemented by the Howard government. As well, the Rudd Labor government had added no new MPAs in its first term of office.

This election farce highlighted the policy failures of the previous five years in marine policy: a return to sectoral management, a drawn-out regional planning process which had produced one Regional Marine Plan in a decade, a loss of community input into marine policy, and a vacuum in terms of institutional arrangements for implementing a now twelve-year-old oceans policy. Clearly the Rudd government had not seen marine policy as a priority, despite the size and importance of the Australian EEZ. Marine policy had been left meandering along inside a government department, the remaining external bodies with an interest in the area (NOAG and MCCN) having been dissolved. This had followed the Howard government's dismantling of the original institutional arrangements it had introduced earlier to implement an integrated and comprehensive oceans (marine) policy.

An unanswered (and possibly unanswerable) question was whether this downgrading of marine policy by successive governments of different persuasions was a case of neglect or a planned deliberate unravelling of arrangements (disintegration) that, without doubt, favoured the dominant sectoral interests (oil and gas mining, commercial and recreational fishing) over those arguing for integrated planning and management (predominantly scientists and conservationists).

Overview of current coastal and marine policy in Australia

The Gillard government's November 2010 response to the recommendations in the latest coastal zone inquiry in Australia (HORSCCCWEA 2009) is consistent with previous federal governments' responses to the past three major national coastal inquiries over the last 30 years. At the establishment of the inquiry, expectations are raised that the federal government would become involved in coastal planning and management. The submissions to the inquiry, in general, recognised the policy failings associated with a minor federal government role, and these submissions are acknowledged in the recommendations of the inquiry, which always proposes a greater federal role and several new agencies to assist in implementation.

The federal government's response, though, when confronted by the reality of this greater role, has been to propose a more modest approach.

This time the coastal policy development process is to be couched in terms of a response to the impacts of climate change on the coast and possibly – with the balance of power in federal parliament held by four independents and one Green (three from coastal seats) – rural development (which could include 'seachange' elements).

Given the potentially very substantial financial costs that will be involved in preparing Australia's coastal infrastructure for the impacts of climate change and the need for a coordinated national response (in terms of planning in particular), it is difficult to see that the primary taxing (revenue-raising) government can avoid a continuing and more direct involvement in implementing coastal policy in the next decade and beyond. Yet to achieve this outcome will mean that this particular government will need a policy intervention that no other federal government has achieved.

The picture for marine policy development and implementation in Australia is an even more tragic one. From the great hopes and expectations that surrounded Australia's ratification of the UN Law of the Sea Convention, and the extraordinary speed with which Australia developed the world's first Oceans Policy that promised a comprehensive and integrated implementation of ocean policy, this area has descended into a sectorally-dominated approach with little interest shown in it by the level of government with sole responsibility - the federal government.

There had been a suggestion that the independent review of the EPBC Act 1999 (Hawke 2009) might have revitalised consideration of marine policy, given the Act's coverage of Commonwealth marine waters as an area of national significance. But whilst marine policy was mentioned in the review, it was as an aside and there was no suggestion of greater action in this area in a revised Act.

The future for Australian coastal and marine policy

Overall, Australia seems to have adopted a position where successive federal governments are not able 'to walk and chew gum at the same time' in terms of marine and coastal policy in Australia; that is, governments of both persuasions have only been able to focus on one area - marine *or* coastal policy development and implementation - at one time.

The Rudd government (2007-2010) seemed to be only able to focus on coastal policy to the exclusion of marine policy, whilst its predecessor Coalition government seemed to be able to focus on marine policy and ignore, for most of its term in government, coastal policy. By the time the Howard government finally turned its mind to coastal policy with the 2006 framework of ICZM, it had begun to lose interest in its own oceans (marine) policy. The Gillard government appears likely to focus on climate change impacts on coastal communities as its contribution to coastal policy development, and has said nothing on marine policy at all.

What is clear over this extended period of almost one quarter of a century is that neither major party in government has embraced integration of marine and coastal policy. So what does the future hold? Is there any way out of the current chaotic policy picture? There were some NGO proposals presented at the last federal election which may influence the new government and help address these policy failings. These are introduced below, incorporated into a brief summary of the separate current challenges facing marine and coastal policy domains.

Australian marine policy

The major environmental policy failure has been a return to the pre-Oceans Policy framework of sectoral control over the policy and processes operating in determining resource allocation in the marine environment. This has led to a resumption of business as usual, based on the mid 1990s model of the power and control over the marine public domain lying squarely with the major corporate stakeholders - for example oil, gas, fishing, shipping - to the almost total exclusion of the larger community. This has been reinforced by the tendency of the Federal Government - removed as it is physically from day-to-day operations in the marine environment – to view stakeholder engagement in terms of engagement with designated sectoral stakeholder lobby groups as described by Wescott and Fitzsimons (2011), rather than with individual members of the wider community. This has been reinforced by the loss of conduits to the wider community (such as the 10,000 registered participants in the defunct Marine and Coastal Community Network), and the splitting of marine policy administration from the coastal policy administration within the federal bureaucracy.

The ACF has offered an alternative approach to this – the implementation of an Oceans Act. The ACF's Act would cover all aspects of the current non-statutory Oceans Policy, including institutional arrangements and regional marine planning. With the assistance of the National Environmental Law Association (NELA), the ACF has actually prepared draft legislation (Smyth, pers. comm.) and is engaged in lobbying the Gillard government to adopt this approach. Whilst these efforts continue, the solution to the current governmental lack of interest (and hence failure) in marine policy in Australia may lie instead in raising the public profile of marine policy. To do this, marine policy will probably have to 'piggy-back' on the much greater public interest in 'the beach', that is, coastal policy. In turn, further development of coastal policy (see below) will probably 'piggy-back' on the Gillard government's commitment to climate change action and possibly the government's (electorally-based) interest in regional affairs.

In the meantime, there is an urgent need for a re-examination of Australia's marine governance arrangements. Possibly an immediate way forward would be for the Greens to instigate a Senate inquiry into marine governance in Australia now that they hold the balance of power in the Senate(from 1 July 2011).

Australian coastal policy

Prior to the last election the Australian Coastal Society (ACS) presented both parties with a three-point plan for coastal policy reform in Australia. They recommended (ACS 2010):

- the development of a national coastal policy by all three tiers of government under an intergovernmental agreement on coasts;
- the establishment of a national coastal information system to ensure a strong data base across environmental, social and economic areas on which to base the policy; and
- the instigation of a national coastal commission under a federal coastal commission act, to drive development of the policy and its implementation.

The government's response to the House of Representatives' report and recommendations proposes using the existing Coasts and Climate Change Council to aid in the development of an intergovernmental agreement on coasts. Hence the most optimistic slant that can be placed on revitalising coastal policy in Australia towards the ACS proposal is the use of this council as the first step towards the progressive adoption of the three points above.

Any resurgence in national coastal policy in Australia appears to be now firmly tied to two major political issues:

1. The impacts of climate change; and
2. The revived dominance of regional and rural politics.

This is, in part, evidenced by the current sole national coastal policy body – the advisory Coasts and Climate Change Council – being as much about climate change as coastal governance and being responsible to the Minster for Climate Change and Energy Efficiency (not the minister covering wider environmental matters).

As well, the rise in importance of marginal regional coastal electorates (approximately half of these 33 seats are still marginal) appears to be significant, with four of the five non-major party members of the House of Representatives from coastal seats (two urban, two regional).

Integrated coastal and marine policy

This chapter began by suggesting that a commitment to integrated ocean and coastal management (including integrated coastal zone management) could be used as a measure of marine and coastal policy success. Clearly, if integration is the aim, then tackling environmental policy failures in the marine and coastal domain requires treating the two domains as inter-linked (and connected completely to catchment management practices as well in a catchment-coast-oceans continuum).

The above analysis shows that there is a continuing lack of integration between national marine and coastal policy, and that this continues to be the major environmental policy failure in this domain. Hence, it is somewhat surprising, and a little disappointing, that even some NGO bodies in the marine and coastal fields do not appear to be recommending integration of marine and coastal policy whilst, to their credit, certainly suggesting ways of addressing policy failures within the coastal and marine policy domains.

But if the core elements are extracted from the NGOs' proposals and combined with those enunciated for an integrated approach in Wescott (2009), there are a series of common elements which might address most of the national marine and coastal policy failures when fully implemented. These proposals are:

1. A national ocean and coastal act;
2. A statutory national marine and coastal strategy;
3. A statutory national marine and coastal council; and
4. A clear marine and coastal resourcing program based on a 'matching fund' formula from federal and state governments. The resourcing program would be based on the strategy above as well as on the principles of integrated ocean and coastal management. The council above would oversee the program.

Conclusion

National marine and coastal policy in Australia appears to be a classic 'victim' of Australia's federated system. Here is an area which splits planning and management of the environment between the jurisdictions of federal and state governments and hence has always lacked a cohesive holistic policy response. The states and territories control the coastal zone out to three nautical miles offshore, and the Federal Government from there out to 200-plus nautical miles offshore.

But an observer might suggest that the challenge to governance emanating from the twin threats of climate change and coastal development should, indeed almost certainly will, require a uniform national response. It is indeed difficult to envisage how state and local governments are going to cope with the financial challenges of adapting to climate change and how Australians will cope with a different planning approach in each state and territory.

What appears to be necessary to address these challenges is the emergence of a political champion, or champions, to both drive coastal policy reform based on the government's response to the 2009 House of Representatives report and to re-energise and completely review Australia's Oceans Policy.

History suggests that a ministerial champion is the key factor in initiating and driving federal government marine and coastal policy reform. Senator Robert Hill ensured an Oceans Policy was developed within two years in the late 1990s, and David Kemp pushed through a sixfold increase in 'no-take' marine zones in the Great Barrier Reef Marine Park in the early years of this century.

The lingering question now is – does such a ministerial champion currently exist, and if one doesn't, will one emerge in the near future?

Chapter 8

Water privatisation: Failure of public policy

MICHAEL BUXTON

Introduction

Nineteenth-century colonial Australian governments vested the ownership and control of water in the state. But from the early 1990s, governments began to adopt market-based principles in legislation and policy for water allocation. The use of market measures to allocate rural public water resources is one of the most fundamental policy shifts in the nation's history. Different values underlie water allocation principles. The approach taken here reinforces the view of Jessica Weir (2009, pp. 18-19) who argued that further change to sustainable uses of rural water will require 'the need for profound change in our intellectual traditions', and that 'change in government policy will not come about through science, economics and management but through fundamental changes in values'. Claims made for the benefits of a market-based approach and the reasons for the shift to this approach have been insufficiently addressed. This chapter reviews these claims and concludes that claimed benefits are yet to be realised, indicating policy failure to this point.

Institution-building

From the late 1800s, irrigation water use expanded through the construction of major storages and systems to transport water across large areas. Allan (2004) calls this the period of industrial modernity. Almost 90 per cent of Australia's water storages were built between 1950 and 1990 (Smith 2003). Over 450 large dams, with a combined capacity of 81,000 gigalitres (GL), have been constructed in Australia, including 200 major storages in the Murray-Darling Basin (Ball *et al.*

2001). Governments bore the cost of construction of dams and irrigation schemes, and for many years irrigators paid far less than the cost of the water, administration and scheme maintenance. Large publicly-funded state institutions oversaw rural water policy, system operation and allocations. Substantial public spending was allocated to private gain as irrigators profited through income and increases in the capital value of farms with tied water allocations and low water costs. The growth of irrigation storages stabilised from the mid 1970s, and of urban water storages from the mid 1990s (see Figure 8.1).

Figure 8.1 Growth of water storages in Australia

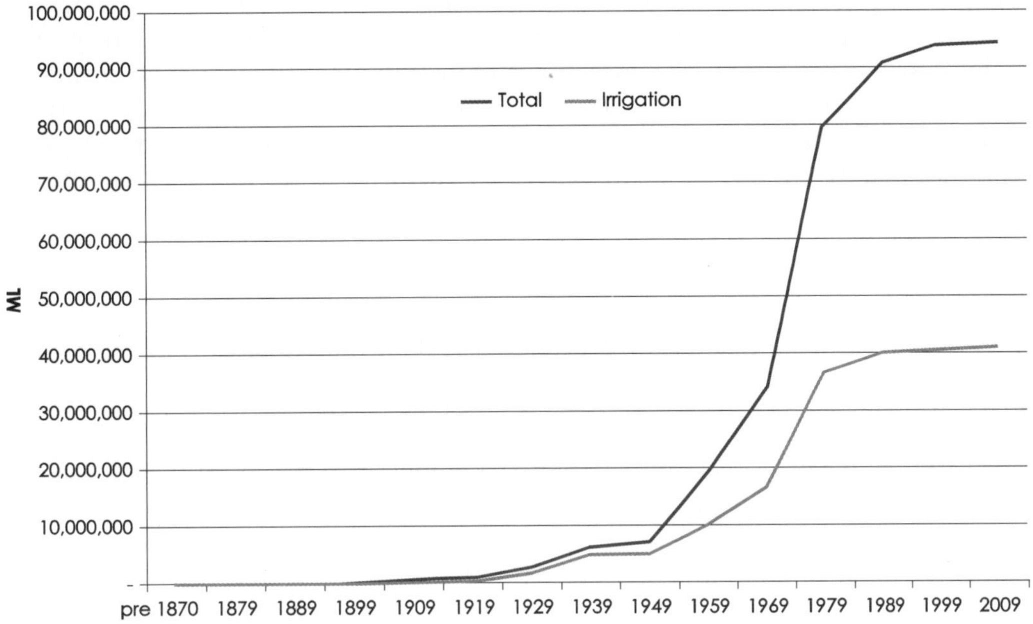

Source: Compiled from ANCOLD 2010.

The development of large water storages in Australia was facilitated by a recasting of water rights in the late 1800s. In Victoria in 1884, Alfred Deakin chaired a Royal Commission into the future of irrigation in the state. Victoria's *Irrigation Act* 1886 implemented the key Royal Commission recommendation that the state take control of Victoria's water and extinguish the old common law riparian rights system (DCE 1991). Later, legislation continued this principle of public ownership, for example, the *Water Act* 1958 vested in the Crown 'the right to the use and flow and to the control' of all water. Water allocations were based on an administrative system of rules rather than the less well-defined system of riparian rights, which gave priority to owners of land adjacent to waterways. Deakin's principles were gradually applied to all Australian states (DWR 1986).

Solving the protracted disputes between the lower Basin states over water sharing was seen to be an early test for the Commonwealth government after Federation. Connell (2007) describes the various conferences, Royal Commissions and other inquiries that followed Federation, ultimately culminating in the 1915 River

Murray Waters Agreement, which was passed into law by the Parliaments of Victoria, New South Wales (NSW), South Australia and the Commonwealth. Essentially, the Agreement provided: a guaranteed minimum flow to South Australia; the sharing of water held in storages between NSW and Victoria, giving those states the right to develop water resources of their River Murray tributaries; and establishment of a commission to oversee the development of the river's water resources (Connell 2007).

The pre-federation colonies' strong attachment to their water resources is reflected in the Australian Constitution. Responsibility for water was not included in those areas for which the Commonwealth Parliament could make laws. Section 100 specifically limited the power of the Commonwealth Government with respect to water. However, a range of Commonwealth Government powers have now been applied to water policy, in particular the section 51 powers which relate to trade and commerce, corporations and external affairs, and section 96, which allows the Commonwealth to make specific-purpose payments to states and to set conditions under which those payments are made. Section 96 was used to provide assistance to states under the National Action Plan for Salinity and Water Quality and the Natural Heritage Trust programs of the late 1990s early 2000s. Section 92 allows the Commonwealth to make laws for trade and commerce, and to extend these laws to navigation and shipping. This allowed the Commonwealth to be party to the 1915 agreement between Victoria, NSW and South Australia for sharing the River Murray waters.

The Commonwealth's section 51 powers have most recently been used to enable the Commonwealth Government to become directly involved in water planning through the establishment of the Murray-Darling Basin Authority, using a combination of Commonwealth constitutional powers and a referral of some powers from the states. This led to the development of an intergovernmental agreement recognising this Commonwealth role and enabling the Commonwealth to prepare and implement the Basin Plan, which will establish sustainable diversion limits (SDL) for each catchment within the Basin and a program for recovering water to achieve the SDL.

A new regime for water

From the 1990s, Australian states began to progressively adopt the view that self-managing water markets, not government regulation, were the most efficient and effective mechanism for allocating the rural water resource. In NSW, for example, this occurred when the *Water Management Act* 2000 replaced the *Water Act* 1912, and in Victoria when the *Water Act* 1989 replaced disparate water legislation. This involved allocating perpetual, tradable property rights to private users of a publicly-owned resource. This policy 'abandoned Alfred Deakin's fundamental principle that irrigators should not hold property rights in water' (Cathcart 2009, p. 254).

The Victorian *Water Act* 1989 illustrates this process. Legally, 'rights' to land and water in Australia are entitlements conferred by legislation. The Victorian act assumed that rights existed through historical use, and that these rights should be converted or 'saved': 'existing entitlements will continue as long as their holders wish' (DWR 1987a, p. 8). As a result, 'a valuable private interest is created by an irrigation water right ... [and] ... The Water Act provides those private rights with strong underpinnings in the form of bulk water entitlements' (DWRb 1987, p. 63). Cathcart (2009, p. 258) has argued that property rights created 'a gift from fellow taxpayers who having handed over their property are now required to 'buy it back' on behalf of the environment'.

The 1989 Victorian legislation and other states' water legislation separated water entitlements from land as a precondition to establishing tradable water property rights. The Victorian legislation defined property rights over water in terms of four principles – that rights were explicit, exclusionary, enforced and transferable:

> *In most fields of activity...[the loss of rights of use]...would be intolerable. Private property is strongly defended and well defined in law just to prevent such an eventuality...the operation of a market economy depends on people being assured that the fruits of their endeavours will not be arbitrarily confiscated. Property rights in real estate, bank accounts, stocks and bonds and so on are all* ***explicit****. Rights to the enjoyment of property are for the most part* ***exclusive*** *to the owner. The rights of ownership can be* ***enforced*** *at law. And the owner normally has the right to* ***transfer*** *property to others. Popular concepts of justice and equity underlie such a system of property rights (DWR 1987b, p. 66; emphasis in original).*

Despite this rhetoric, private property rights over water differ from those over land. State legislation provides that the Crown owns water resources. For example, section 7 of the Victorian *Water Act* 1989 still contains this principle: '7 Continuation of the Crown's rights to water (1) The Crown has the right to the use, flow and control of all water in a waterway and all groundwater'. The Crown allocates water and can intervene to adjust the size of the available resource under defined conditions and through periodic reviews. Entitlements are shares of the available resource and licences to use or trade shares are affected by climate and other factors. This approach assumed that individual and corporate profit-seeking would lead to superior outcomes than public interest motivation.

Sturgess and Wright (1993) have argued that the practice of prior appropriation – under which the appropriators of a natural resource are accorded a priority, tradable right for no cost – was applied to water trading in Australia because of the increasing scarcity of the resource. Scarcity is determined by both extraction and inflow. The willingness by governments to agree to new intergovernmental arrangements has been influenced by the persistent drought affecting much of southern Australia. In Victoria, for example, most of the state received average or below average rainfall between 1997 and June 2009. This translated to reductions in stream flows for all regions ranging from twenty per cent in the north east to over 80 per cent in the western part of the state (DSE 2010). Murray-Darling storages were also significantly affected.

However, governments showed a commitment to a market ideology long before the impacts of extended drought became clear. Both this commitment and political considerations resulted in alternative policy measures being rejected, such as auctioning allocations, land retirement, or identifying regions where irrigation is sustainable and withdrawing allocations from environmentally degraded farmed areas. The two motivations of markets and politics were linked. Governments were reluctant to withhold water from severely salinity-affected areas even by compensating landowners, or to mandate a transfer of allocations from low- to high-value uses. In times of drought, the market was seen as the most efficient means of reallocation. The alternative under the former regulatory system would have been for governments to create a hierarchy of uses.

Complexity and fragmentation

Australia's federal system of government has complicated the development of national water policy because of separate state water laws, varying management systems, and perceptions of state interests. Parties to Murray-Darling Basin agreements have been described as 'rival nations' (Alexandra & Eyre 1993). Intergovernmental policy arrangements are yet to achieve substantial improvements. The Murray-Darling Basin Commission was established in 1988, replacing prior arrangements. The Commission operated on a consensus basis and developed policy, operated river systems and managed assets while states developed their own water legislation and management systems within the framework of intergovernmental water sharing agreements. It took several years for the three Murray-Darling states to agree, in 1989, on a salinity and drainage strategy, and for the Commonwealth, state and territory governments to agree to a cap on Basin water extractions so that diversions were limited to the 1993-94 level of development. Groundwater has not been properly included in extraction limits, although limits are now being considered (NWC 2010).

The 1992 Murray-Darling Basin Agreement put in place an intergovernmental agreement for the Basin. In 1994, the Council of Australian Governments (COAG) adopted a strategic framework for the reform of the Australian water industry, the *National Water Reform Framework*. Priority was given to the establishment of water markets. In 2004, the Intergovernmental Agreement on a National Water Initiative (NWI) was adopted by all governments as the national framework for change to the water sector. The National Water Commission was established to oversee implementation of the NWI, but states, territories and ministerial councils are used to implement it. The NWI constitutes the overriding national policy framework for the rural water sector.

However, the NWI also seeks to achieve a broader range of economic, social and environmental outcomes including the return of over-allocated systems to environmentally sustainable levels of extraction, and integrated catchment management. It also seeks to facilitate more efficient market mechanisms including lower transaction costs, removal of other trading barriers, and the establishment of reliable water registers which better specify access. The NWI attempts to use water trading to improve environmental use. These objectives are still to be achieved, with the NWI finding that consistent water pricing is not in place, environmental allocations are not agreed, and slow progress is being made on environmental water accounting (NWC 2009).

The NWI, legislative and policy objectives and principles, catchment management processes and other Basin agreements are delivered by statutory and non-statutory water plans. These plans are the means for allocating water and delivering management processes. However, the development of water plans is 'now critically inadequate with over 40 per cent of plans yet to commence' (NWC 2009, pp. 37). The different processes and institutional arrangements and processes have resulted in varying types of plan. Gentle and Olszac (2007) have argued that water planning has avoided resolution of the major issue of balancing consumptive uses and water for the environment with the result that decisions on major reallocations have been deferred. They argue that the lessons from other industry reform associated with the cessation of industry assistance have not been learned and, consequently, that adjustment packages and compensation have become major political issues.

Under the Australian Constitution, governments are required to pay just compensation for the resumption of property, but continued vesting of water in the Crown limits property rights. The process of repurchase of water from irrigators is being implemented in the Murray-Darling Basin under the Commonwealth's Water for the Future Program. By December 2010, almost 1,000 GL had been purchased. The 2002 Living Murray initiative from the Murray-Darling Basin Ministerial Council defined a healthy river as 'a compromise, agreed to by the community, between the condition of the river and the level of human use' (MDBMC 2002, p. 47). Connell (2007) describes the business-as-usual approach, which prevented radical change. Between 1,900 and 4,000 GL of water needed to be returned to the river, but the Living Murray program proposed only three alternatives, 350 GL, 750 GL or 1,500 GL. Eventually, 500 GL was to be returned through water efficiency measures, again postponing the issue of reallocation (MDBMC 2005).

Some governments remain reluctant to purchase water for environmental purposes. The Victorian Government has rarely purchased water, preferring savings through transfers, infrastructure and other efficiencies, reflecting multiple objectives. The Commonwealth Government has committed $3.1 billion to purchase water entitlements for the environment. The draft Murray-Darling Basin Plan being prepared under the Commonwealth *Water Act* 2007 proposes a reduction of 20-22 per cent or 3,000-4,000 GL in allocations to users. The government is still considering the political implications of this proposal.

The move to a market-based system

Critics of regulated water systems used three main arguments for the adoption of market-based principles: first, that regulated regimes had failed because of the inadequate financial performance of water authorities and because of over-allocation of a scare resource; second, that the use of markets would lead to more efficient and responsible use; and third, that water users needed security for investment. These claims will now be examined, and water policy discussed including policy in largely non-irrigated peri-urban regions of Australian cities.

Failure of regulated systems

Regulated water systems, where the authority responsible for allocating the water was also the authority responsible for selling the water, certainly led to over-allocation and environmental degradation. Private property rights and privatisation were seen by market advocates as ways of achieving more profitable financial performance. Privatisation can take many forms. Despite their legal ambiguity, water property rights attempt to convert a public resource into a private asset. Privatisation of state water authorities represents a further step away from the nineteenth century and modernist models, while a corporate public sector model has also been applied to Australian water authorities. Privatisation in other countries, particularly the United Kingdom (UK) and France, was used as an argument for the same policy response in Australia. However, Johnson (1993, p. 164) has shown that UK privatisation was carried out:

> *...at a massive financial cost, with gains that did not justify it. The necessary arrangements to achieve the privatisation objective cost British taxpayers far more than the resources generated from the sale of the water companies and did not introduce the 'market disciplines' that were supported to ensure its future efficiency and effectiveness.... These developments call into question ... the drive to corporatise and privatise some of Australia's large water authorities....*

Walker (1993) has argued that pre-privatisation earnings from assets and sales compared favourably with publicly listed corporations. Public water authorities made high returns to government with significant public benefits. Walker argued for the need for agencies such as the Industry Commission to improve financial accounting procedures to avoid the allegation that findings are developed to support preconceived ideological positions.

The overriding question to be asked about more recent water policy in Australia is: has the establishment of market-based systems led to public benefit without substantial environmental and economic costs? In recent decades, irrigated agriculture has traditionally used between 75-77 per cent of the approximately 24,000 GL of water used in Australia each year (NLWRA 2006). Yet Australian gross farm production was valued in 2006 at just $17.6 billion, or about three per cent of GDP, with one per cent of agricultural land irrigated producing 28 per cent of agricultural production (ABS 2006). Between the years when the establishment of the Murray-Darling Basin Ministerial Council was first proposed (1984-85) and the introduction of the Basin cap (1996-97), average water use for irrigation increased by between 65-76 per cent (Shadwick 2002; Edwards 2003; NLWRA 2001).

Almost 90 per cent of water used in the Murray-Darling Basin is used for irrigation (Ball *et al.* 2001; ABS 1992). By the mid-1990s, about 80 per cent of the Murray River's average annual flow had been diverted for irrigation (Alexandra & Riddington 2004). Total water use in Australia increased substantially after the establishment of intergovernmental arrangements designed to manage and control water use and extractions. Large increases in surface and ground water extractions and in irrigation use correlate strongly with the establishment of the Murray-Darling Basin Agreement in 1982, the Ministerial Council in 1986, and the Murray-Darling Commission in 1988. This process continued under a market-based system, as Bell and Quiggin (2008, p. 713) note: 'Thus far, the activation of water markets appears to have increased not decreased demand for extractive water usage'. Cork and Delaney (2005) have shown that:

> *Total water use in Australia increased by 65% between 1984 and 1997: surface water use by 58 % (mainly in NSW, Qld and Vic), groundwater use by 88 % (mainly in WA, NSW and Qld), and water use for irrigation by 76 % [Seckler et al. 1998; van Hofwegen & Svendsen 2000; Dunlop et al. 2002] … water extraction is either close to or above sustainable levels in 26 % of surface-water basins (mainly in the Murray-Darling Basin) and 30 % of groundwater management units (mainly in the Lake Eyre Basin) [NLWRA 2001].*

The introduction of tradeable water entitlements coincided with a substantial expansion of irrigated land, by 22 per cent or 450,000 hectares between 1993-94 and 2000-01 (ABS 2004). Expansion for cotton and sugar in northern NSW and Queensland contributed to this trend.

Environmental impacts of over-allocation remain serious. By 2003, Murray-Darling Basin allocations exceeded the cap by almost 20 per cent (Young & McColl, cited in Wahlquist 2008). The National Competition Council assessment of Victoria's water reforms shows that flow levels remained inadequate in 43 of the 73 systems where bulk entitlements conversion had occurred. Only in four river systems (5.5 per cent) have environmental flows improved under bulk entitlements. The volume of water entering NSW from Queensland has reduced by over 50 per cent in the twenty years to 2003, partially through on-farm dam construction.

The activation of 'sleeper' (water rights which have not been used) and 'dozer' (rights not currently used) licences are examples of contradictory policy outcomes. The sleeper licence issue is explained as a policy failure by market proponents, but has been integral to a market-based approach, leading to a worsening of the problem of over-allocation and environmental degradation. Gardner and Bowmer (2007, p. 50) point out that 'the prevalent rule in the US prior appropriation states is that only water rights that have historically been used may be transferred'. In contrast, in NSW and Victoria, the two states with the greatest irrigation use, old water licences under respective state water acts were converted to tradable entitlements under new water laws regardless of a history of use. Much of the early trade in water was in sleeper allocations, for example, 99 per cent in an interstate pilot trade project (CSIRO 2000). They constituted about 30 per cent of licensed water allocations in the Murray-Darling Basin (Bell & Quiggin, 2008), and in the late 1990s the activation of sleeper licences had the potential to increase water use in the Murray-Darling by 15 per cent. In some states, such as NSW, sleeper licences to groundwater have also been traded.

More efficient use

States claimed that a system of transferable water property rights 'would eventually promote water use in more efficient and economically viable enterprises' because 'irrigators in marginal situations would have a negotiable asset, namely their water entitlement' (DWR 1986, p. 20). Trade in water entitlements is generally limited to trade between irrigators, initially within local or regional districts, but is now expanding to interstate trade. The 1994 COAG agreement and the 2004 NWI policies continued the attempt to introduce a water trading system which would lead to more efficient use of water through allocation to higher-value uses. In 1989-90, at the beginning of the introduction of the market-based system, 50.6 per cent of irrigation water was used for pasture, 34.7 per cent for cropping and only 14.7 per cent for horticulture (ABS 1992). The historic subsidy to irrigators encouraged low-value agricultural activities. Most water used for irrigation continues to produce relatively small returns per hectare under the market-based system. For cropping, NSW rice growers used about 2,000 GL of water to produce a crop valued at $350 million in 2001-02. By 2003, irrigated pastures, wheat and rice fields and sugar plantations used about 70 per cent of irrigation water, whereas high value products such as fruits, vegetables and cotton used only about 30 per cent. About two-thirds of the value of irrigated crops came from about one-third of water used (ABS 2004).

Trade may occur in seasonal and permanent allocations. Peterson *et al.* (2004) have shown that most trade has been in seasonal allocations, but even so, in a study of three main irrigation districts between 2000 and 2003, only 11-20 per cent of annual allocations and only 1-2 per cent in permanent allocations were traded. Just 3.4 per cent of farms traded water in 2003-04 (Productivity Commission 2006), although since then even a number of Victorian districts have reached the 4 per cent figure of permanent allocations traded. The Victorian government report *The Value of Water* (DNRE 2001) outlined the growth of trade in Victoria over the 1990s. The temporary market was by far the largest, with between 100,000 megalitres (ML) and 250,000 ML per year being traded by the latter half of the 1990s. Approximately 25,000 ML was traded annually on the permanent water market. The Victorian government's annual water trade reports illustrate the importance of water trading as a way of managing through the drought. The 2008-09 trade report reveals several developments with respect to water trading: interstate trade; water for new irrigation development; a greater presence of urban authorities in the market; and new risk management options provided by carryover in Victoria (DSE 2010).

Increased efficiency can raise the demand for the available resource, as Bell and Quiggin (2008) point out. This increased demand has been met mainly by savings from infrastructure improvements, with half the water saved under the NWI $6 billion plan to be given to irrigators. Trade can lead to the transfer of water from healthy to degraded river systems, or from degraded to new irrigated areas. Nevertheless, trade is not leading to extensive reallocation of water to higher-value uses, although there is evidence of some trade going to new irrigation systems that complied with environmental best practices (O'Neill 2004).

Interstate water trade was a major goal of the 1994 reforms, but physical and policy restrictions exist between interstate trade, limiting it in volume and extent: trading rules became complex; water is lost in transmission and this needed to be accounted for; different states had different policies regarding reliability of allocation; some states allowed back trade (a process that allowed trade to occur upstream so long as an equivalent trade downstream has occurred); and new administrative procedures were required to track the trades and accommodate water moving into and out of valleys where diversions had been capped by the 1997 MDB Ministerial Council decision. Water pricing reforms are now resulting in charges moving towards meeting headwater charges, system maintenance (though not sunk costs), and contributions towards environmental costs.

Secure access for investment

Two of the ten objectives of the NWI are the development of secure water access entitlements and the progressive removal of barriers to trade in water. The NWI argued that:

> *Where access to water is insecure, users are likely to lack the necessary confidence to invest in new capital equipment, better management and infrastructure … The NWI established a framework for developing clear, nationally compatible and secure water access entitlements, to be defined in stature as a perpetual or ongoing and exclusive entitlement to a share of water (NWC 2009, p.5).*

Advocates for the market-based system of tradeable water property rights all argue for the need for durable property rights for water in order to achieve further investment, particularly in high-value uses. However, the limited extent of trading, particularly of permanent entitlements, the continuation of water over-allocations and environmental degradation, and the lack of extensive transfer into higher-value uses all point to the limited results of a market-based system of secure private entitlements.

Peri-urban water policy and use

In Australia urban water supply and storage facilities are often located in the peri-urban interface (Bunker and Houston 2003). For example, the majority of water for urban and irrigation use in South East Queensland is supplied from 53 storages and three major flood control dams in the peri-urban region (SEQWater 2002b). In Victoria, all major storage reservoirs with the exception of the Thomson Reservoir are located in the Melbourne peri-urban region (Melbourne Water 2005). The water supply needs of cities lead to extensive impacts on both the peri-urban hinterland and broader rural areas.

In addition to being sites for supply and storage of water, peri-urban regions are locations of varying types of consumptive and non-consumptive water uses. Much of Australia's horticulture is produced in the peri-urban areas of capital cities on high-quality soils unserviced by state irrigation systems. Water is extracted from streams in peri-urban areas for bulk entitlements (defined shares of water from specified sources), and through licences as private diversions from unregulated rivers. Considerable ground water is also extracted, and farm dams store water which would otherwise flow into streams.

Demand for water in peri-urban areas is increasing, but development is not being adequately integrated with water availability or cost. Populations are increasing in most Australian peri-urban areas at urban rates of increase (Buxton *et al.* 2008). In a number of Victorian urban supply systems, for example, sustainable yield limits will be reached well before 2055, some as early as 2010-14. This situation worsens under different climate change scenarios.

Generally throughout peri-urban Australia, a substantial decline in rainfall occurred over the ten years to 2009. Water inflows fell substantially with severe impacts on storage levels. For the Melbourne peri-urban area, for example, by 2030 there is the possibility of a 30 per cent reduction in run-off in the south-east, and between five per cent and 45 per cent in the south-west because of climate change (CSIRO 2000). Table 8.1 compares Campaspe Basin water storage levels between 2004-08, illustrating the impact of the ten-year drought to that year.

Table 8.1 Major storages in the Campaspe Basin

	Capacity at FSL* (ML)	Current Volume (ML)	Current % Full
Eppalock	312,000	19,800	6.3%
Upper Coliban	37,500	191	0.5%
Lauriston	19,780	7,772	39.3%
Malmsbury	17,780	328	1.8%
Sub-Total	387,060	28,091	7.3%

* Full supply level.

Source: Department of Sustainability and Environment 2008, *Monthly Water Report*, June, DSE, Melbourne.

In catchments in the Melbourne peri-urban area, environmental allocations have borne a disproportionate reduction in water availability compared to reductions in other uses as supply has fallen with the reduction in long-term average stream flows. For example, in the Campaspe Basin between 2004-06, total water available fell from the long-term average of 305,000 ML to 104,460 ML, the total used fell from 95,100 ML to 69,620 ML, but the catchment outflows fell from 185,100 ML to only 10,040 ML (see Table 8.2). Water resource commitments for catchments are very high, and these are contributing to a substantial reduction in environmental flows.

Table 8.2 Campaspe Basin surface water balance

	Long Term Average	Average over 2004-05 & 2005-06
Catchment Inflow (ML)	305,000	101,300
Inter-basin transfers and water stored or released from storages (ML)		3,160
Total available	305,000	104,460
Bulk Entitlement (ML)	64,740	39,770
Unregulated River Private Diversions (ML)	1,560	1,850
Farm Dams (ML)	28,800	28,000
Total Used	95,100	69,620
Losses (ML)	24,800	24,800
Catchment Outflows (ML)	185,100	10,040

Sources: Department of Sustainability and Environment 2006, *State Water Report 2004-0 – A Statement of Victorian Water Resources*, DSE, Melbourne; Department of Sustainability and Environment 2007, *State Water Report 2005-06 – A Statement of Victorian Water Resources*, DSE, Melbourne.

Groundwater is a valuable source of water with licence entitlements across Victoria totalling 853,000 ML. Peri-urban irrigation use is relatively low and mainly for pasture and vegetables. Stock and domestic water can be sourced from groundwater without a licence (although a bore construction licence is required), accounting for 19 per cent of total groundwater use in 2004-05. This unlicenced extraction could become a more serious problem, with increasing population and declining stream flows.

Spatial land fragmentation is leading to increased construction of farm dams, reducing water supply in streams. Victoria has over 300,000 farm dams (Lake & Bond 2007). In some catchments, stock and domestic farm dams represent a substantial share of the overall water use within that catchment, for example, at 47 per cent in South Gippsland, 35 per cent in Maribyrnong and 30 per cent in Otway Coast. Of the 4,181 houses constructed in five outer peri-urban municipalities between 1997 and 2007, 3,103 were constructed on lots two hectares or larger and making extensive use of farm dams.

Land use change, particularly the conversion of agricultural land to plantations and other high-water-using crops, can reduce surface run-off and groundwater recharge and, in some places, draw water from aquifers. The timber industry, for example, continues to affect water yields from catchments from both logging and incremental plantation establishment. High levels of water are consumed by regrowth trees, in comparison to mature forest stands (Wilby & Gell 1994). In the

Glenelg catchment in western Victoria, plantations account for 39 per cent of the sustainable yield of catchments and 8.3 per cent of the surface run-off (SKM, CSIRO & BRS 2010). A paradox for water planners is that, to combat climate change, large-scale tree planting for carbon sequestration is advocated and this in turn can lead to a reduction in water yields from catchments.

Sectoral and cross-sectoral policy approaches

The belief in the ability of self-regulating markets to rationally allocate water is a form of neo-liberal governance, or the attempt to reduce the role of the state. It is likely that both a water market and regulatory measures can make important contributions towards solving problems of rural water use, but only if the limitations of both and reasons for their past failures are recognised.

An important reason for the failure of regulatory systems was sectoral and institutional fragmentation. Water policy and management became regarded as the proper and necessary province of water specialists. Other government sectors were little involved in water matters, particularly land use agencies, with the result that sectoral decision-making prevailed and water use was never fully integrated with land use and other relevant sectoral areas.

A sector is a field of activity which defines knowledge according to disciplinary concepts and methods. Sectors tend to divide and separate knowledge and activity. A sectoral agency 'tends to be exclusive and inward looking dedicated to maintaining limited and exclusive goals, and to perpetuating its organisation culture against attacks or criticism from other sectors or broader interests' (Buxton *et al.* 2006, p. 244). Australian water agencies tended to consider water policy as a limited and independent, not as an integrated, activity. Water agencies also developed close associations with industries benefiting from allocations, such as irrigation interests. Musgrave (in Mourtiz 1992) has argued that regulatory institutions which administered irrigation schemes were captured by irrigation users.

In contrast, cross-sectoral policy requires that sectoral agencies consider the conceptual and methodological approaches of other sectors in framing proposals, and consider the reciprocal impacts of proposals. All successful intergovernmental agreements require the partial surrendering of sovereignty to shared decision-making bodies. The potentially most effective international water agreements are cross-sectoral, although even these are hampered by perceptions of the national interest by participating governments. Intergovernmental arrangements over water for the last three decades have never satisfactorily addressed the vertical institutional fragmentation between local, regional, state and national bodies. Nor have these arrangements ever satisfactorily addressed the horizontal fragmentation between state agencies and even within multi-functional state agencies that underlies the water governance system. Institutional and sectoral fragmentation has thus prevented the development of integrated policy responses to existing and emerging problems.

Conclusion

Claims made for market-based principles are either ideological or empirical. Ideological claims can be identified as motives for change, while empirical claims can be tested. Supporters of a market-based system merged values, ideological motivations and empirical claims. Three decades after state and national political leaders recognised a range of serious issues associated with the systems of water allocations, major problems remain. Environmental water has been regarded as a competitor to human use instead of fundamental to the maintenance of river health and ecologies (Sinclair 2001). Peter Cullen's comment that Murray-Darling Basin water development 'is a history of articulate interest groups seeking … particular advantage' (cited in Wahlquist 2001, p. 166) has yet to be fully addressed. While governments are increasingly moving away from terms such as 'water rights' and are adopting terms such as water shares, for many the concept of a 'water right' prevails and there is an expectation that a given quantity of water will be delivered for consumptive use.

The release of the guide to the proposed Murray-Darling water plan (MDBA 2010) has polarised debate about recognising and responding to the environment's need for water. The government response to the plan will test the resolve of governments to implement the concept of water shares. The problems of misallocation and over-allocation of water, failures in water accounting, and pricing inadequacies – all characteristic of former regulated systems – largely remain to be rectified. The belief that the environmental costs of water allocations would be addressed by market-based allocations to higher- value uses remains unfulfilled. For the present, the result is continuing policy failure.

Much effort is being expended in state and intergovernmental processes to assist the efficient operation of markets to achieve objectives, including the further public purchase of water rights. The success or failure of these endeavours during the next few years will allow a more precise estimation of whether market-based systems have led to public benefit without undue environmental and economic costs.

The contribution of Danny O'Neill and Katelyn Samson to the preparation of this paper is acknowledged, although the views expressed are those of the author.

Chapter 9

Climate adaptation and the Australian system of government: The Gold Coast example

MICHAEL HOWES AND AYŞIN DEDEKORKUT-HOWES

Introduction

For more than two decades scientists have been warning governments at all levels about the serious risks posed by climate change. For the last few years the media have prominently covered the issue, and there has been a consistently strong public demand for action. Despite this, successive governments have struggled to develop and implement responses, as climate sceptics and powerful industries delayed action. Two factors are major contributors to the situation. First, climate change is by nature a 'wicked' problem, and so is not readily amenable to conventional policy or planning solutions. Second, the governing institutions of Australia emerged in the late nineteenth century and were simply not designed to deal with such issues. Each factor will be dealt with in turn below. The current situation on Queensland's Gold Coast offers a useful case study of climate adaptation policies and plans. Exploring the difficult decisions that need to be taken here generates three proposed strategies for action that may enable the three levels of government to collaborate more effectively. Policy success in this area could take the form of improvements in the resilience and reduced vulnerability of settlements like the Gold Coast. A lack of action that leaves such places at high risk from the impacts of climate change would be a textbook policy failure.

The nature of the problem

Although the science of climate change is complex, the underlying principle is relatively simple. The devil, of course, is in the detail. Since the industrial revolution, human activity has released large quantities of greenhouse gases (such as carbon dioxide) that have changed global climate patterns. The environmental consequences include: higher average temperatures; rising sea levels; the loss of snow and ice cover; shifting patterns of precipitation; the accelerated loss of habitats and biodiversity; increased land degradation; and more frequent and intense extreme weather events (Solomon *et al.* 2007; CSIRO & Bureau of Meteorology 2010; Arndt, Baringer & Johnson 2010; Australian Academy of Science 2010). These changes will, in turn, have significant social, economic and political consequences, including: disruptions to agricultural production; exacerbated freshwater scarcity; increased damage to built assets; higher insurance costs and defensive expenditures; the spread of vector-borne diseases into more populous zones; a rise in heat-related deaths; a substantial increase in refugee numbers; and disputes over access to water (Stern 2006; Parry *et al.* 2007; UNEP 2007; DCC 2009; Australian Academy of Science 2010).

Governments have considered several policy and planning responses, and a few have been recently adopted in a piecemeal fashion. These responses can be grouped into two main categories. The first consists of mitigation actions designed to cut net greenhouse gas emissions so as to reduce the speed and scale of climate change. The second consists of adaptation responses designed to manage the impacts that cannot be avoided (DCC 2010; Garnaut 2008; Metz *et al.* 2007; COAG 2007). Mitigation policies have been dealt with extensively elsewhere, and are *palliative* in their impact. Adaptation has been less extensively explored, but offers longer-term *transformative* options. Mitigation and adaptation are obviously connected; put simply: the less we mitigate, the more we have to adapt (Kane & Shogren 2000; Stern 2006).

One of the standard texts used in training public sector policy-makers is *The Australian Policy Handbook* (Althaus, Bridgman & Davis 2007). It argues that policy-making can be understood as a cycle of eight steps: issue identification; policy analysis; policy instruments; consultation; coordination; decision; implementation; and evaluation. The authors themselves, however, admit that the process is not always so straightforward, as this ideal model and the whole cycle grinds to a halt when faced with 'wicked' policy problems. The concept of 'wicked' problems emerged from an article on planning theory by Rittel and Webber (1973), who assigned them ten attributes:

1. They are difficult to define;
2. There is no clear end or boundary to the problem;
3. There are no agreed criteria to judge the correctness of a response;
4. Responses have unforeseen consequences;
5. Responses that go wrong cannot be easily undone;
6. It is not possible to identify all possible options;
7. There is no suitable precedent to guide decision-makers;
8. They are interconnected with other problems; and,

9. There is no agreed explanation of the problem.
10. Mistakes in either action or inaction are very costly.

In a nutshell, these attributes make understanding a 'wicked' problem as well as developing, implementing and evaluating an effective response highly problematic.

As a policy problem, climate change has all of these attributes (Head 2008). In terms of climate adaptation, tens of thousands of scientists around the world have been studying the problem for several decades, but it has not been possible so far to accurately predict the impacts on a specific city let alone a particular property. The impacts that have been identified are presented as a range of possibilities under several different scenarios (Dessai & Hulme 2004; Parry *et al.* 2007). Climate change is a process that operates over centuries, which is well beyond the normal *policy-making* horizon, and has no clear end or boundary. A range of adaptation responses is possible (e.g. restrictions on new developments, the construction of flood barriers, or the abandonment of some areas), but there are no agreed criteria by which the options can be judged. Any attempts to adapt will have long-term environmental, economic and social consequences that cannot easily be reversed. There is no easy solution, but options may emerge from unforeseen technical innovations or changes to settlement patterns.

Climate change is largely the product of other problems (such as the depletion of fossil fuels, air pollution and land clearing), and will itself exacerbate existing issues such as the loss of biodiversity, water scarcity, falling food production, increased disease, and rising demands on the public purse. The causes, nature and seriousness of the problem are each publicly disputed by climate sceptics, making action difficult. Finally, a great deal is at risk: failure to act will be costly, but taking the wrong path will also have costs. These attributes have led Ross Garnaut (2008), who led the Australian Government's climate change review, to argue that climate change is not just 'wicked' but diabolical.

This 'wicked' nature has fed the politics of climate scepticism, which encourages official neglect of the problem. The Howard Coalition Government (1996-2007), for example, was often accused of doing too little (Christoff 2005; Mercer & Marden 2006; Head 2008). Prime Minister Howard himself openly admitted that he was 'sceptical about a lot of the more gloomy predictions' of climate change research (Brissenden 2006; Pearse 2007, 2010). Others within the Liberal party were even more forthright. Senator Nick Minchin famously asserted that the issue was actually a conspiracy by left-wing activists seeking to attack capitalism (Ferguson 2009). Minchin was later instrumental in ousting Malcolm Turnbull from the Liberal leadership because Turnbull supported the Rudd government's Carbon Pollution Reduction Scheme. His replacement, Tony Abbott, is more sceptical and once described climate change as 'crap' because he believed the science was uncertain (Grattan 2009; Pearce 2009). Linked to this scepticism is the entrenched opposition of high-emitting industries that dominate the Australian economy and its politics. Research by Pearse (2007, 2009), for example, suggests that the resources sector, and the coal industry in particular, exerted undue influence over the Howard government's climate change policies. Nor did this influence stop with the change of government at the 2007 federal election. When the Labor government tried to impose a new tax on the resources sector in 2010, for example, the industry response was a major factor in the removal of Rudd as Prime Minister. The new leader, Julia Gillard, moved quickly to appease the sector, with major concessions and policy changes (Mitchell 2010). This had led some people to question the limits of Australian democracy (Menadue 2010). All of this indicates that the problem is so 'wicked' as to put the Australian political system under extreme stress. To understand why, and to find a strategy for adapting to climate change, we need to understand the underlying architecture of the Australian system of government.

The nature of Australian governments

In his seminal work, *The Risk Society*, Ulrich Beck (1992) pointed out that governing institutions designed in the nineteenth century have difficulty responding to complex twentieth-century risks. While this observation was derived from his study of European society, it applies equally well to Australian governing institutions facing twenty-first century problems like climate change. It must be remembered that this system was developed in a very different technological, economic, social, political and environmental context (Howes 2005). Most of the fossil fuel-based production, transport, housing, entertainment and communication technologies upon which we now rely had either not been invented or were not in common use. The economy relied mainly on agricultural production, and the large trans-national firms that now dominate most sectors were nowhere to be seen. Socially, the population was much smaller and more dispersed, most of the social safety nets like welfare benefits and Medicare did not exist, and the level of education was lower. The system was developed decades before all but one of the current Australian political parties came into existence (the exception being the Labor party), before the concentration of economic power in the hands of a few business leaders, before the rise of ubiquitous and instantaneous electronic media coverage, and before the rise of new social movements (for example environmentalism and human rights groups). Finally, the system predated much of our scientific knowledge about the environment and the rise of global environmental issues like climate change.

The Australian political system is the product of a nineteenth-century compromise between the governments of six colonies. Each had enjoyed considerable autonomy, and their leaders were reluctant to hand over powers to a new national government. The first constitutional convention in 1890 failed, and it was another seven years before an agreement was reached at the second convention. The resulting system of government was later to be described as the 'Washminster mutation' because it combined elements from the US government in Washington and the Westminster system in Britain (Jaensch 1997; Thompson 1980). As in Britain, the monarch was the head of state, and the executive (prime minster and cabinet) were members of parliament. The new Australian legal and administrative systems, too, were much the same as the British. From Washington came a federal political system (with each colony becoming a state), a written constitution, a high court to interpret and defend the constitution, and a senate with equal representation for the states (Howes 2005).

In terms of the distribution of powers, the Constitution specified a restricted set of responsibilities for the new federal government. All other residual powers fell to the states. Since environmental problems had not been recognised as significant policy issues at the time of federation, and thus not recognised in the Constitution as federal responsibilities, the states have consequently taken the lead role (Toyne 1994). Two territories, the Australian Capital Territory and the Northern Territory, were established and later given limited self-rule, but the federal government retained the power to override their decisions. No mention is made in the Constitution of local government, so councils exist entirely at the mercy of the states. This means that they can be amalgamated, replaced or abolished at will. Struggles between all three levels of government have been a central feature of Australian politics ever since.

Over the last century, successive federal governments have found ways to adjust their powers using three main strategies: delegation, co-operation, and intervention (Howes 2005). Delegation occurs when the federal government chooses not to be involved in a particular issue and is happy to leave it to the states. Passive delegation occurs when it does not attempt to enter a policy area (for example, most air pollution laws are left to the states). Active delegation effectively licenses the states to act on the Commonwealth's behalf (this happens, for example, with environmental impact assessments under the *Environment Protection and Biodiversity Conservation Act* 1999). There are also bilateral funding agreements: federal funding in return for federal say over a state issue.

Co-operation, or co-operative federalism as it is often known, is perhaps the most common strategy. Here the federal, state and territory governments (and sometimes representatives from local government) establish a joint decision-making body to agree on a common set of goals and to coordinate enforcement across all jurisdictions. The 1992 Intergovernmental Agreement on the Environment, for example, led to the creation of the joint ministerial National Environment Protection Council (NEPC). Complementary legislation was passed around the country so that measures agreed upon by the NEPC automatically become enforceable in all jurisdictions (Howes 2005).

The intervention approach is where the federal government expands its powers to take control of an issue. This can occur in several ways (Toyne 1994; Howes 2005). First, it can use an existing power in a new way. In 1983, for example, the Federal Government used the external affairs power given to it by the Constitution to stop the Franklin Dam in Tasmania. It successfully argued in the High Court that the area in which the dam was going to be built was protected under the World Heritage Convention and that the Constitution gave the Federal Government the power to enforce such treaties. The second strategy is to persuade the states to hand over or refer some of their powers to the national government. This happened during World War II when the states surrendered some of their taxation powers, and more recently when they handed over the power to regulate corporations. Finally, the federal government can use its fiscal powers to fund programs that by-pass the states altogether. A good example is Landcare, where funding for environmental rehabilitation is provided to community organisations directly (Howes 2008; Howes 2005).

The three strategies of the federal government are mirrored at other levels. In terms of delegation, state governments often leave minor environment and planning issues to local councils. However, co-operative processes can also handle more substantive issues. The *South East Queensland Regional Plan 2009-31*, for example, was largely a state government initiative (in which the Department of Infrastructure and Planning [DIP] took the lead role), but the process did include consultation with the affected local councils and they in turn use the plan to guide their decision-making (DIP 2009). In terms of intervention, the Queensland government maintains a central control of local development through its direction of funding in the *SEQ Infrastructure Plan and Program 2010-31* (DIP 2010).

Local councils can sometimes develop innovations that are well in advance of the other levels of government (Adams & Hine 1999), and they tend to be wary of state government policies and plans (Burton & Dredge 2007). The Council of Mayors SEQ (South East Queensland), for example, is an advocacy group that lobbies state and federal governments for policy changes. In 2008, the Queensland government unilaterally exercised its powers to force many councils to amalgamate against their will. Even when state governments adopt delegation or co-operative approaches, councils are often suspicious of cost shifting, where they are given more responsibility to implement policies but the necessary funding is not provided (Liebrecht & Howes 2006). Power struggles, turf wars and the lack of trust all undermine relations between different levels of government and their agencies. This is one of the main barriers to effective collaboration, and stymies attempts to get well-coordinated responses to issues that cut across jurisdictions (Howes 2008; Liebrecht & Howes 2006).

Given this situation, it is no surprise that the overall response to climate adaptation has been slow and complex. In terms of delegation, the federal government has left most of the detailed planning work to the state, territory and local governments. In Queensland, for example, the State Government released a *Draft SEQ Climate Change Management Plan 2009-31*, and the Gold Coast City Council (GCCC) adopted its own *Climate Change Strategy 2009-14*. With regards to co-operation, the Council of Australian Governments (COAG) agreed to a strategy on adaptation that required collaboration between all levels of government (COAG 2007). On the intervention front, the Federal Government created the Department of Climate Change, and directly funded various programs (DCC 2010). Similarly, the Queensland Government established the Office of Climate Change in 2007 for 'coordination, implementation and review of Queensland's climate change strategies and whole-of-government policy recommendations for achieving further cuts to the state's emissions profile' (Queensland Government 2010).

The fact that such a complex range of strategies, policies and plans slowly emerged across various levels of government is the legacy of decisions taken in the late nineteenth century about the design of the Australian political system. Further, the lessons from the climate mitigation side of the policy equation do not bode well. By the end of 2007, for example, all state and territory governments, as well as both major parties at the federal level, supported a greenhouse gas emissions trading scheme. By 2010, however, the consensus had been destroyed by short-term politics, political rivalries, and a persistent campaign of opposition by sceptics supported by a powerful fossil fuel sector. These forces combined to play on the weaknesses within the existing political system and, as a result, the legislation was blocked in the Senate and was eventually deferred. The idea was revived by the Gillard government's Clean Energy Future legislation that was subjected to a hostile parliamentary debate in 2011, but even if passed the policy is under threat of being rescinded if there is a change of government. So, how well has this system done when it comes to helping one of its most vulnerable cities, the Gold Coast, adapt to the impacts of climate change? What can we learn from these past policy failures? What strategies would predispose climate adaptation policies and plans to success?

Case study: The Gold Coast

If climate change, by definition, constitutes a wicked policy problem, the Gold Coast provides a highly sensitive case study due to its inherent physical vulnerability and limited social resilience. In physical terms, the region is dominated by low-lying, flood-prone land surrounded by an exposed coastline and numerous waterways (GCCC 2009; DCC 2009; Parry *et al.* 2007). Added to this, development has taken place in vulnerable areas, with waterfront property being much sought after. The present population of the region is around 500,000, and this is predicted to rise to 900,000 by 2030 (GCCC 2006). Such rapid population growth puts pressure on existing infrastructure and encourages further development in low-lying coastal areas that will be particularly vulnerable to floods, storm surges and rising sea levels (GCCC 2009).

In terms of social resilience, the Gold Coast has a large retirement community, with 20.9 per cent of the population aged 60 or more (GCCC 2006). Older Australians are the most at risk from heat stress and heat-related deaths from higher temperatures. They will also be more susceptible to vector-borne diseases that may spread to the region (Baum *et al.* 2010). Over 24 per cent of the Gold Coast's population was born overseas, and 3.8 per cent of the total population does not speak English very well (GCCC 2006). This will make any community engagement difficult, as many people may not understand the nature of the threat or what they can do to minimise the risks. Finally, although there are a considerable number of very wealthy people living on the coast who can afford to take individual adaptation measures, 17.5 per cent of households are classed as low income (GCCC 2006). The lack of resources reduces their ability to pay for defensive measures or to move to less vulnerable dwellings.

There have been numerous policies and plans from all levels of government with significant climate adaptation elements that will affect the Gold Coast directly [see Table 9.1 for a selection and Dedekorkut *et al.* (2010) for further detail]. As it is not possible to give a detailed analysis of them all, this chapter will focus on some key initiatives that are indicative of the strategies outlined in the previous section. This in turn should signify the capacity of the Australian political system to respond to the challenge of climate change adaptation.

Table 9.1 Selected policies and plans relevant to climate adaptation on the Gold Coast

Level	Policy or plan
National	1. National Climate Change Adaptation Framework, April 2007, Council of Australian Governments. 2. Adapting to Climate Change in Australia: An Australian Government Position Paper, 2010, Department of Climate Change. 3. National Climate Change Adaptation Research Plan: Settlements and Infrastructure, December 2010, National Climate Change Adaptation Research Facility.
State: Queensland	4. Toward Q2: Tomorrow's Queensland, 2008, Department of the Premier and Cabinet. 5. Queensland Coastal Plan, 2011, Department of Environment and Resource Management. 6. ClimateQ: Toward a Greener Queensland, 2009, Department of Environment and Resource Management. 7. ClimateSmart 2050, 2007, (including ClimateSmart Adaptation 2007-12), Environment Protection Agency (now Department of Environment and Resource Management).
Regional: South East Queensland	8. South East Queensland Regional Plan 2009-2031, 2009, Department of Infrastructure and Planning. 9. South East Queensland Infrastructure Plan and Program 2010-2031, July 2010, Department of Infrastructure and Planning. 10. South East Queensland Natural Resource Management Plan 2009-2031, 2009, Department of Environment and Resource Management. 11. Rural Futures Strategy for South East Queensland 2009, 2009, Department of Infrastructure and Planning.

Table 9.1 continued

Level	Policy or plan
	12. Draft Connecting SEQ 2031: An Integrated Regional Transport Plan for South East Queensland, 2010, Department of Transport and Main Roads. 13. South East Queensland Water Strategy, 2010, Queensland Water Commission. 14. The South East Queensland Healthy Waterways Strategy 2007-2012, SEQ Healthy Waterways Partnership. 15. Draft SEQ Climate Change Management Plan, 2009, Queensland Department of Infrastructure and Planning.
Local: Gold Coast	16. Gold Coast Planning Scheme, 2003 (amended January 2007), Gold Coast City Council. 17. Gold Coast Climate Change Strategy 2009-2014, 2009, Gold Coast City Council.

The *National Climate Change Adaptation Framework* (COAG 2007) is a good example of co-operative federalism. Like most policies and plans, the problem and impacts are reiterated in summary form to stress the importance of the agreement. The goal is stated concisely:

> *The long term goal of this Framework is to position Australia to reduce the risks of climate change impacts and realise any opportunities. In the medium term (5-7 years), targeted strategies in this Framework will build our capacity to deal with climate change impacts and reduce vulnerability in key sectors and regions (COAG 2007, p. 4).*

Ministerial councils are mentioned as a key mechanism for coordinating actions and implementing policies and plans. Two areas for action are identified: 'building understanding and adaptive capacity' and 'reducing vulnerability of key sectors and regions' (COAG 2007, 6). In terms of the first area, the National Climate Change Adaptation Research Facility (NCCARF) is meant to play a key role, and the main focus of the agreement is to support research that will inform decision-makers with regard to impacts and vulnerability. With regard to the second area, priorities are identified as: water resources; coastal regions; biodiversity; agriculture, fisheries and forestry; human health; tourism; settlements, infrastructure and planning; and natural disaster management. Each has a section outlining the need for research to provide a basis for future decision-making and actions.

At the national level, the federal government recently released *Adapting to Climate Change in Australia: An Australian Government Position Paper* (DCC 2010). The paper restates the nature of the problem and summarises some of the key impacts that are unavoidable. It acknowledges that the Federal Government does not have the capacity to solve the problem alone, and it argues that responsibility for adaptation is shared by all levels of government, business and the community. Government in general is cast as a capacity builder and reformer that will help business and the community adapt by providing information and by setting the right conditions through polices and plans. The government also commits to changing the way it manages programs and assets. While the Federal Government plans to play a leadership role, state and territory governments are seen as having the primary task of taking direct adaptation action because they are the main providers of public goods and services (for example, hospitals, schools, parks, roads, water, sewerage, and garbage collection). Priority areas for action are identified as coastal management, water supply, infrastructure protection, maintaining natural systems, disaster management and assisting agriculture. The paper advocates the embedding of climate change adaptation considerations in the policy-making process. It points out that $12.9 billion has been allocated to the Water for the Future program, and reminds us that NCCARF was created in order to improve information for governments, business and the community. Here we can see all three strategies of the Federal Government at work. In terms of intervention, it funds its own programs and protects environmental assets of national importance. In terms of co-operation, there is a focus on working through COAG on various schemes. Finally, in terms of delegation, there is an acknowledgement that much of the heavy lifting will be left to other levels of government, as well as the business and community sectors.

ClimateQ: Toward a Greener Queensland (DERM 2009) is Queensland's policy framework for transitioning to a more sustainable and climate-friendly economy. The strategy summarises climate change impacts on Queensland's regions, previews the economic and social costs, and presents eight sectoral strategies: energy; business; planning and building; community; primary industries; transport; ecosystems and government. Three of the key themes identified have adaptation elements: investing in the productive future of key industries, conserving significant ecosystems, and adapting to the impacts of climate change. *ClimateQ* includes investments and policies totalling $196 million, including $87 million in new initiatives. In terms of specific adaptation actions, *ClimateQ* focuses on the areas of disaster management, water use and biodiversity protection. The strategy identifies the role of the state government as providing information to assist communities and businesses in preparing for the risks of climate change. Most of the initiatives in the strategy focus on research and capacity building. The CSIRO climate flagship research project is conducting a comprehensive assessment of climate change risks in South Eastern Queensland (SEQ), and the data is informing regional planning efforts. There is $8 million to be invested in developing a digital elevation model of the state to better understand the impacts of sea level rise, storm surge and coastal erosion. The resulting interactive maps will be available to local governments and other stakeholders for future land use planning. The state government also provides funding for various programs (for example, the Water Smart Buildings Program, Rural Water Use Efficiency Initiative, climate change corridors for biodiversity, and disaster management education programs) and prepares the general guidelines under which local governments operate. The government will also ensure that infrastructure grants to local governments and communities take account of both greenhouse gas reductions and climate change adaptation considerations.

The *SEQ Draft Climate Change Management Plan* (CCMP) (Queensland Government 2009) contains thirty two proposed actions for state and local government to support the implementation of the climate change policies of the *South East Queensland Regional Plan 2009-2031* (DIP 2009). Thirteen of these actions are concerned with natural hazards and climate change adaptation, the rest are mitigation. Seven of the thirteen adaptive actions propose to review and revise existing plans, policies and guidelines, develop new guidelines and strategies, or implement state policies through lower level planning. The remaining six adaptive actions focus on data collection, synthesis and communication. Although the CCMP does not provide detail of how these actions will be implemented, it suggests that finding practical and cost-effective ways to make a real difference is possible by working collaboratively with other levels of government, the development industry, councils and the residents of South East Queensland. Like the *SEQ Regional Plan*, the draft CCMP was prepared by the state DIP in consultation with the affected local councils.

Finally, at the local level, there is the Gold Coast City Council *Climate Change Strategy 2009-14*, which again deals with both mitigation and adaptation (GCCC 2009). As with all the polices and plans considered, it starts with a summary of the problem based on the research to date, then moves on to acknowledge the vulnerability of the Gold Coast region. Further research, raising public awareness and making infrastructure resilient, all appear in the seven strategic outcomes. There are also calls for coordinated actions within the council and between the different levels of government. In terms of implementation, most of the specific actions listed consist of reviewing and auditing current operations or further research. While much of the strategy could be internally funded, there is a call for more external funding.

The overall effect of all these policies and plans for the Gold Coast is that the first tentative steps towards adaptation have been taken, but a lot more needs to be done. With regards to water supply and flooding, for example, a series of studies were commissioned by DCC, the GCCC and the Queensland government. A desalination plant has been built to supplement water supplies during droughts, and both the local council and state government have run effective water conservation campaigns. The local Hinze dam has been connected to a grid that allows water to be transferred between the major reservoirs within the wider region during dry spells. The dam wall has been raised to increase storage capacity during dry periods and reduce the risk of flooding from overflows during wet periods. More detailed flood maps were released to the public in 2009, and sea level rises are now being taken into account in planning decisions for new developments. There is still, however, the problem of what to do with existing built assets that are at risk, and how to stand up to politically powerful developers who still want to build on vulnerable land. The area already faces some major problems, driven by its rapidly growing population and variable climate, notably limited water supplies, storm damage, land degradation and coastal erosion.

Where to from here?

The first thing to acknowledge is that some tough decisions need to be made about how best to adapt to the impacts of climate change. Given that the public purse is limited, who should be given financial assistance to adapt and who should miss out? Which properties are going to be defended and which will be abandoned? How much money should be allocated to adaptation, and who will pay? Where does the legal liability lie when properties lose their value or lucrative developments are stopped (England 2007)? (In the case of the Gold Coast, section 704 of the Queensland *Sustainable Planning Act* 2009 requires local councils to compensate owners when a planning decision reduces their property's value.) These decisions are all the more difficult given the limited public resources, competing demands from other policy areas, and the power struggles between different levels of government. Such conditions create a real risk of policy failure.

The second point is that the current economic and political context is going to discourage any individual government from taking steps that might be unpopular. At the national level, political opponents could block adaptation measures in the Senate, and then use the issue to hound people from office. We have already seen this happen with the watered-down mining tax. Further, any action by one level of government might be resisted and undermined by the other levels. State recalcitrance on reforms for the management of the Murray-Darling system and the funding of public hospitals are prime examples. Finally, although local governments have made a start, they do not have the resources or power to implement all the necessary changes, and are wary of state governments trying to shift costs onto them without appropriate funding.

The problem will not get any easier to address as time goes on, and it is unlikely that we will see a major restructuring of the political and economic system in the foreseeable future. So, what can be done? Most of the climate change policy research to date has focussed on the mitigation side of the equation (Bulkeley & Betsill 2005; Christoff 2005; Mercer & Marden 2006; Pearce 2007; Head 2008; Bulkeley & Newell 2010), although this is starting to change (Byrne *et al.* 2009; Head 2009). We suggest that there is a set of three complementary strategies that together might accelerate the implementation of adaptation policies and plans:

1. Using the established NEPC processes and institutions;
2. Creating a new climate program modelled on Landcare; and
3. Establishing a new national statutory authority.

This set of strategies gives the Federal Government a key role because: it has already indicated a desire to take the lead, it is best placed to withstand local resistance to individual decisions, and it commands the largest pool of useful resources. These strategies are compared in Table 9.2.

Table 9.2 Strategies for accelerating adaptation actions

	Risk of an individual government becoming the target for public or business anger	Risk of being blocked by the Senate	Risk of state government resistance
1. New NEPC measure on climate adaptation	Low, responsibility diffused between all levels of government	Low, requiring either a creative interpretation of the existing act or a minor amendment	Low, a collective decision
2. New climate program within Caring for Our Country	Low, onus put back on community & business	Minimal, Landcare-style programs are popular	Low, funding flows from national to local levels
3. New climate change adaptation statutory authority	Low, blame deflected to new authority	Medium, major parties disagree on federal intervention but Greens hold balance of power	Medium, states balance dislike of federal intervention against wish to avoid blame

The first strategy would be to use the existing NEPC structure to create a new National Environment Protection Measure on Climate Adaptation. Such a measure would be enforceable in every jurisdiction under existing state and federal legislation, although it would require either a creative interpretation of the areas defined for action in section 14(1) of the *Commonwealth NEPC Act* 1994 or a minor amendment. This would, however, reduce the risk of any one government becoming a target for blame as the decision is taken collectively. The second strategy is to use the Landcare approach, where the Federal Government offers funds for climate adaptation projects proposed by various combinations of community groups, businesses and government agencies at any level. Climatecare could be added to the current Caring for Our Country suite of programs that generally have support from all sides of politics. While the majority of Landcare projects have traditionally been rural, some have been undertaken in more urban areas such as the Gold Coast. Such a move would diffuse blame by putting the onus back onto the local community and business, which would have to identify the major problems in their area and implement solutions. The third strategy is to create a new 'arms-length' National Climate Change Adaptation Statutory Authority and then delegate to it the power to make unpopular decisions. This approach has been used successfully in the past. The power to set interest rates was handed over to the Reserve Bank, for example, in a move supported by both sides of politics. Raising rates is always going to be unpopular with business and the community because it imposes significant costs, but good monetary policy requires this action during times of high inflation. While the new authority is obviously not going to be in the same league as the Reserve Bank, the same principle could be applied, but the legal basis would have to rely on either a referral of powers from the states or the use of the external affairs power in enforcing the 1992 UN Framework Convention on Climate Change. This strategy would face the most resistance from the Liberal-National coalition in the Senate, but the Greens, who hold the balance of power (from July 2011) may be persuaded to support such a move.

These three strategies would work together to support each other and would require substantial funding from the Federal Government to avoid the problem of cost shifting. Perhaps some currently allocated resources could be redirected from closely-related programs like Infrastructure Australia. It must be conceded that all three strategies proposed are band-aid measures, but they would at least speed up the deployment of tangible actions on climate adaptation. A more substantive change would involve a major restructure of the government, business and community sectors, which is well beyond the scope of this paper. (There is a long history of literature in this area, see for example Dryzek (1987) or Eckersley (2004).)

So, what would these three strategies mean for the Gold Coast? The National Environment Protection Measure on Climate Adaptation would be legally enforceable by the Queensland Government and would require the co-operation of the Gold Coast City Council (GCCC) in a similar fashion to the *SEQ Regional Plan*. In order to avoid cost shifting, the Climatecare program would enable the GCCC to apply for Federal Government funds to undertake the necessary capital works. Local firms and/or community organisations might also apply for funding, either in partnership with the GCCC or independently. Finally, both the Queensland government and the GCCC could refer the hardest decisions to the National Climate Change Adaptation Statutory Authority to deflect potentially damaging local political conflicts towards a more removed arm of government.

Conclusions

Climate change poses a 'wicked' problem for the Australian political system and has led to fierce power struggles inside, between and outside governments. These have already contributed to the removal of both a prime minister and a leader of the opposition, hobbled major climate change policies, and exposed the political power of resource industry leaders. Meanwhile, the sceptics have effectively hijacked the public debate. In these circumstances, policy failure will be marked by a lack of action, leaving vulnerable communities at risk of suffering from the impacts of climate change (for example, the continued development of low-lying coastal land that is prone to inundation). A system of government designed as a political compromise in the nineteenth century is struggling to deal with this situation, and the difficulty of future adaptation decisions that have to be taken will discourage action at all levels of government. Policy success will be marked by an increased resilience and reduced vulnerability to climate change. In the short term there are some strategies that would help, but in the longer term the systematic and institutional weaknesses exposed by the issue of climate change need to be addressed.

Notes and acknowledgements

An earlier version of this paper was presented to the Australian Political Studies Association Annual Conference 2010. The authors would like to thank the reviewers and editors who provided valuable feedback. The Urban Research Program and a Griffith University Research Grant funded part of the research on which this paper is based.

Chapter 10

Catastrophic bushfire, politics and the public interest

DAVID MERCER

Disasters expose our social structure and culture more sharply than other important events. They reveal starkly the failure of organizations, regulations, and the political system. But we regard disasters as exceptional events, and after a disaster we shore up our defences and try to improve our responses while leaving the target in place (Perrow 2007, p. 3).

Introduction

Death, injury and asset damage statistics following natural disasters are a relatively simple, shorthand means of rapidly gauging policy 'success' or 'failure' in relation to the management of such events at many different levels. Invariably, alleged policy failings are blamed upon 'unprecedented' events, or 'residual risk', that could not have been anticipated or planned for. This is an argument that has been mounted in recent times in the United States (US), for example, to absolve the relevant authorities from blame in the cases of serious oil pollution in the Gulf of Mexico and the devastation of New Orleans as a consequence of Hurricane Katrina (Congleton 2006).

But there is a growing body of literature that forensically compares and analyses similar disasters in different places, and highlights institutional failings and poorly co-ordinated governments and agencies as a direct cause of excessive fatalities and damages bills in some areas by comparison with others (Burton 2010). Both Florida and Cuba, for example, have a much better record than Louisiana in terms of mitigating the worst effects of severe hurricanes. Similarly, on 04 September 2010, an earthquake of magnitude 7.1 rocked the city of

Christchurch in New Zealand. While there was some building damage, there were no deaths and only minor injuries. Five months earlier, an identical magnitude earthquake hit Qinghai Province in China, killing 2,187 people, injuring 12,000 and rendering 100,000 homeless. After statistically controlling for population, these examples highlight the importance of strict building code regulations, in particular.

Here, we focus on the question of policy 'failure' with particular attention on the bushfires that hit Victoria on 07 February 2009, resulting in the deaths of 173 people and the loss of some 2,100 dwellings. Given that this particular disaster was the subject of exhaustive scrutiny by a subsequent Royal Commission, there is a rich body of documentation of the manifold policy failings involved and recommendations for change. Ongoing legal cases now before the courts are also highlighting claims and counter-claims of culpability and failure.

'Risk' and the state

Over time, in all advanced liberal democracies, governments have steadily extended their regulatory reach over all aspects of the safety, security and health of citizens. Invariably, change has been hastened in response to a particularly pressing problem or specific disaster, and as a precautionary measure for managing future risk (Vlek 2010). Initially, in the nineteenth century, for example, state intervention in the provision of infrastructure for water and sanitation in what are now called 'developed' cities had a dramatic and immediate impact on the numbers dying from cholera and typhoid each year. Sweeping reforms in areas as diverse as policing, medical care, housing, land use planning, education and air quality quickly followed and had equally positive consequences for people's health, security and life expectancy. Parallel developments occurred in the insurance sector. In its modern form, house insurance was triggered as early as 1666 by the Great Fire of London, and saw specialist fire insurance companies opening up in the 1680s, and in the US fifty years later.

Such reforms, of course, are never uncontested. In nineteenth-century England, poverty was widely considered a 'moral failing', and humanitarian philanthropists were obliged to step in to provide housing and other services that were deliberately ignored by the state until much later. Debates around the morality of private insurance have followed a similar path: should broad scale 'safety nets' be made available to all, or only to those who can afford to pay? In Australia, by comparison with many other countries, there is only limited availability of insurance against natural disasters. As we shall see, there are also different models operating between states. Furthermore, controversy initially surrounded the introduction of compulsory seat-belt legislation and child vaccination, as well as the phasing-out of unleaded petrol and the fluoridation of water supplies. More recently, legislation to curb cigarette advertising, as well as the alleged security 'benefits' of internet censorship and the widespread deployment of CCTV (close-circuit television) cameras in public places are far from universally accepted.

In all such instances, conflict is played out between advocates of the 'public interest' on the one hand and libertarians privileging individual rights and responsibilities on the other. Risk expert Professor John Adams (2009) at University College, London, for example, has long been an outspoken critic of the so-called 'nanny state' and a supporter of the view that in recent times many societies and nations have moved much too far in the 'risk averse' direction. This criticism often carries particular weight in public debates around preferred responses to 'low probability' yet 'high impact' events. In his recent Reith Lecture – 'Scientific Horizons' – Lord Martin Rees (2010), for example, identified the banning of 'beef on the bone' as an unwarranted overreaction to the 'mad cow' disease outbreak in the United Kingdom (UK) in the 1980s.

Natural disasters

Public policies relating to the many facets of climate change – especially those associated with 'violent weather patterns and resulting catastrophic events' – present classic examples of this tension, as highlighted in a recent, hard-hitting report on the security aspects of natural disasters by the Australian Strategic Policy Institute (Yates & Bergin 2009). The document appeared seven years after a group of high-level officials presented its report to the Council of Australian Governments (COAG) on *Natural Disasters in Australia* (COAG 2002) and five years after the COAG's more focused *National Inquiry on Bushfire Mitigation and Management* (Ellis, Kanowski & Whelan 2004), which was precipitated in large part by the fires that bore down with such ferocity on the nation's capital in January 2003.

The latter report is significant because, for the first time, there is official recognition that climate change will almost certainly result in more frequent and severe fire events, especially across southern Australia, and that bushfires are consequently more than just an issue for individual state or local government responsibility. The report also contained a major recommendation relating to land use policy that was reiterated with equal conviction in July 2010 in the final report of the Victorian Bushfires Royal Commission.

The post-disaster response traditionally is conceptualised as a threefold process, moving from *relief* through *recovery* to *rebuilding* (or *reconstruction*). The first of these phases obviously involves immediate emergency assistance of various kinds, while the third entails much longer-term strategic assessment and planning. Ideally, this final phase should involve serious questioning of the appropriateness of, for example, rebuilding communities in highly vulnerable locations such as the coast, floodplains, earthquake-prone areas, and the like. In this context, it is also worth pointing out that, for some years now there has been a strong human rights movement coming out of South America, in particular, focused around the 'right to risk management' (Audefroy 2010).

Invariably, as with recent major fires in Australia, the Commonwealth Government has stepped in at short notice to provide much-needed financial and other emergency assistance, as well as funding ongoing, long-term research into climate change and fire behaviour. But for at least one directly impacted interest group – the Fire Brigade Union of New South Wales – this represents a totally inadequate reactive response. Deeply concerned by the growing risk of death and injury to its employees, it commissioned its own recent report on climate change and associated bushfire scenarios which added to the voices recommending urgent action at the Commonwealth Government level to reduce Australia's greenhouse gas emissions (Greenpeace Australia 2010). In passing it is worth noting that this report concluded that, if greenhouse gas emissions are allowed to continue unchecked, the projections for 2050 for the northern edge of Melbourne are for the 'catastrophic' fire danger rating to shift from its current one-in-33 year frequency level to one-in-three.

Fire and the Australian environment

The US doyen of world fire history, geography and policy – Stephen J Pyne (2009) – recently reminded readers that fire has been a feature of planet Earth since the Devonian period, but that: 'The advent of anthropogenic fire was a revolutionary event in Earth history'. Whilst lightning strikes are still a common cause of 'natural' fire starts in many countries, numerous human 'triggers' have also augmented this form of ignition. Going back thousands of years, for example, Australian Aborigines were but one of many indigenous societies around the world that employed some variant of 'fire-stick' farming to 'clean' country and flush out wildlife. Indeed, writers such as Peter Latz (2007) have argued that the arrival of humans on the Australian continent heralded a fundamental, human-induced change in the country's vegetation from forests to savannah grasslands. More recent, anthropogenic triggers include deliberate arson attacks (resulting in many fire-affected areas being immediately declared potential 'crime scenes'), errant fuel-reduction burns or campfires, and sparks from machinery or overhead electricity power lines.

With the exception of the extensive arid areas over much of the Northern Territory, Western Australia and South Australia, all of Australia (around 270 million hectares) is fire-prone. Moreover, fires are a common occurrence with Victoria, alone, witnessing some 500 fires on average each year (Clode 2010). The vast majority of these are rapidly contained and extinguished, but occasionally – as in February 2009 – a fatal combination of extreme weather conditions, a breakdown of communication systems and multiple institutional failings results in a series of individual fires rapidly coalescing in an uncontrollable conflagration.

As noted, in the light of recently released climate change data, major wildfires are likely to become a much more common phenomenon in Australia, as elsewhere in the world (Bowman *et al.* 2009; Lucas 2009). The *State of the Climate* report, produced by the CSIRO and Bureau of Meteorology in March 2010 has highlighted the following:

- The whole country has experienced warming over the past 50 years of the order of 1.5-2 degrees Celsius.
- 2000 to 2009 was the warmest decade on record.

- Second only to 2007, 2009 was Victoria's hottest year on record.
- The number of days with record hot temperatures has progressively increased.
- Rainfall over southern Australia has decreased over the last 50 years.

These drying trends are expected to continue. Indeed, Barnett and O'Neill (2010) point out that Melbourne has experienced below-average rainfall every year since 1996. The 'fire season' is getting progressively longer, with days of extreme fire danger now being possible at any time between October and May. Projections put the likely increase in the number of extreme fire days in southern Australia at anywhere between 15 and 25 per cent under a 'low emissions growth' scenario by 2050, and at 120 to 230 per cent under a 'high' growth scenario. These figures highlight the need for a thorough review of existing policies relating to adaptation to a radically changed fire regime.

The 'normality' of bushfire and increasing vulnerability

Even though bushfires are, by their very nature, episodic and have claimed far fewer lives over the years in Australia than floods, heatwaves and cyclones (Haynes *et al.*2010), when they do erupt with force the loss of life and property over a brief period can be high. Prior to the catastrophic events of February 2009 ('Black Saturday') in Victoria, for example, major bushfires had devastated that state on nine separate occasions (1851, 1939, 1944, 1969, 1977, 1983, 2003, 2005 and 2006). Some of these (for example 1851, 1939 and 2003) affected extremely large areas. The 1851 fires consumed around a quarter of the state (approximately five million hectares) but resulted in the loss of only twelve lives. The 1939 ('Black Friday') fires affected about half this area, but 76 lives and 650 dwellings were lost. The 2003 fires burnt for 59 days within a 2000 km perimeter but miraculously with only one life lost.

These three major disasters were all in parts of Victoria that, at the time, were generally thinly populated. Nevertheless, stock losses were massive (one million sheep in 1851 and 11,000 head of stock in 2003), and in the 1939 fire an entire township (Narbethong to the east of Melbourne) was destroyed. The same township was struck again in 2009. By contrast, the February 16, 1983 'Ash Wednesday' fire on the eastern fringe of Melbourne affected a relatively small area. Yet because of the density of the population, 47 people died, 21 in the township of Upper Beaconsfield alone. 'Ash Wednesday' was a portent of what was in store for peri-urban Melbourne in February 2009 and, in hindsight, should have been the catalyst for many of the sweeping reforms recommended by the recent Bushfires Royal Commission.

Invariably, in Australia, individual bushfire events are described as 'unprecedented'. But with the environmental historian Danielle Clode (2009), and Ron McLeod (2003), the convenor of the inquiry into the Canberra fires of that year, I would argue that the historical pattern of successive fires just outlined makes this an inappropriate description. Major fires have always been a feature of the Australian ecology and landscape, and this certainly will not change. But, unfortunately, there appears to have been little improvement over the years in terms of what Australians have learnt about living with fire, especially over the period since 1939. One of the main reasons for this has to be the aforementioned episodic nature of catastrophic bushfire events and the resultant conclusion on the part of many affected residents that 'it won't happen again, at least in my lifetime' (Eriksen & Gill 2010). Another reason is the entrenched short-term thinking so common a feature of Australian state governments preoccupied with the timeframe of the election cycle. In his submission to the recent Bushfires Royal Commission in Victoria, fire historian Tom Griffiths (2009, p.1) could barely hide his frustration:

> *We know this phenomenon, we know the specific contours of the event, and we even know how people live and how people die. The climate change scenario is frightening. But even worse is the knowledge that we still have not come to terms with what we have already experienced.*

Most importantly, driven by cheap land and housing, the 'tree change' phenomenon and state governments intent on promoting 'housing affordability' as an electoral vote-winner, what has eventuated in recent times has been an exponential increase in population in moderate-to-high, 'at-risk' areas, especially around the edges of the major metropolitan areas (Cottrell 2005; Eriksen & Gill 2010). In the North American literature, such areas are widely referred to as the 'wildland-urban interface' (or WUI). Drawing on the work of the demographer Bernard Salt, Kissane (2010) notes that in recent times 10,000 people per year have been moving into 'tree change' communities within 150 kilometres of Melbourne and that, with 523,000 residents, this now represents ten per cent of the state's population in a relatively small area.

Over the decade 1997-2007, government approval was granted for the construction of 4,181 dwellings in the rural zones of just five municipalities in the fire-prone region on Melbourne's fringe. Over half these properties were on less than eight hectares (Buxton *et al.* 2010). Moreover, in July 2010, just days before the release of the Royal Commission's final report, and with the support of the opposition, planning laws were changed in the Victorian Parliament to allow an additional 43,600 hectares of land on Melbourne's fringes to be rezoned for 134,000 new homes. This is merely the latest amendment to the urban growth boundary, justified by the Planning Minister as being 'fundamental to maintaining housing affordability' (Dowling 2010).

It should then come as no surprise to learn that, as the most densely populated state, Victoria has recorded by far the highest number of bushfire-related deaths in the country (53.6 per cent of the national total for the period 1900-2008 (Haynes *et al.* 2010)), as well as the highest economic losses. As Clode (2009 p. 6) puts it:

> *...every year there are more and more people living in high-risk areas... Every year there are more and more fires being lit...As fast as we improve our capacity to fight fires, we seem to increase our vulnerability to them.*

This reinforces Perrow's (2007, p. 1) central argument in *The Next Catastrophe* that, in addition to the *prevention* and *emergency responses,* there is a pressing need to 'reduce the size of vulnerable targets'. Over the years there has come about a growing expectation that the state will take action to minimise the danger to people and property in high fire-risk areas through such mechanisms as appropriate zoning and building regulations, awareness-raising programmes, well-resourced fire-fighting and policing services, and state-of-the-art warning systems.

Post-fire inquiries

In Australia, 'residual risk' – its perception and management – has always been debated openly and passionately in the course of post-fire inquiries and court cases. Since the first Royal Commission in 1939 (the Stretton Inquiry) into the disastrous fires of that year, Victoria alone has had no less than ten such official inquiries (Dwyer & Esnouf 2008), of which the most recent is the 2009 Bushfires Royal Commission. Over the course of eighteen months, and at a cost of $40 million, the Commission received 1,700 written submissions, held 26 community consultations, and heard from 434 witnesses, generating over 20,000 pages of transcripts. With commendable alacrity, after only six months it presented its interim report, with 51 recommendations, in August of that year (Teague, McLeod & Pascoe 2009).

The final four-volume report, with 67 recommendations, appeared in July 2010, only four months prior to the Victorian election, and resulted in the Premier openly apologising to the people of Victoria for 'system failures' both on the day and in the years preceding the tragedy (Austin 2010a; Teague, McLeod & Pascoe 2010). In the case of both the 1939 and 2009 commissions, the focus was on gathering expert and lay evidence as to the causes, response and lessons that could be learned to prevent a similar loss of lives and assets in the future. In the famous words of Judge Leonard Stretton, who presided over the 1939 Royal Commission: 'We have still not lived long enough'. By this he was referring to the pitiful lack of environmental knowledge and understanding displayed by European settlers in fire-prone Australia (Griffiths 2009).

The precise relationship between commissions and the policy process is not always apparent (Prasser 2006; Rowe and McAllister, 2006). A common feature, as Prasser (2010, p. 14) emphasises, is that 'They inevitably report to the very government … that is often at the heart of the crisis they have been assessing.' What is clear about recent events in Victoria is that the Premier, John Brumby, and his Cabinet had a strained relationship with the 2009 Royal Commission right from the start, fuelled in no small measure by the Commission's relentless attack on the alleged leadership failings of Russell Rees, the Chief of the Country Fire Authority, and of Christine Nixon, the Chief of Victoria Police, on the day of the fire. Even though both of these people have subsequently resigned, they were strongly supported – and in Rees' case re-endorsed – by the Premier during the inquiry process. The recently defeated Labor government had long displayed open antagonism to independent voices in the form of statutory authorities, auditors general and commissions of inquiry that were beyond its direct control (Bachelard 2010a; Coffey, Fitzsimons & Gormly 2010).

Indeed, in the days leading up to the release of the Royal Commission's final recommendations, Premier Brumby made it quite clear that the State Government would not act upon the Commission's findings until his Government had consulted with the community to assess which recommendations were 'reasonable' and which would be rejected or only acted upon at a much later date. By contrast, the then opposition leader – Ted Baillieu – as well as leading fire scientists such as Melbourne University's Kevin Tolhurst, was vocal in urging the Government to accept and act upon all 67 recommendations without delay. Subsequently, when Baillieu became Premier, he announced that his Government supported 59 of the recommendations but would 'consult widely' on the most controversial eight. These included potentially expensive land buy-back proposals, the scrapping of the Fire Services Levy, and the replacement and 'undergrounding' of degraded electricity power lines deemed responsible for five of the fires on Black Saturday. Sensing growing community opposition in the lead-up to the November election, the Premier eventually announced that the Government would act upon all of the recommendations except those relating to land buy-backs and the burial of power lines.

Following the release of the recommendations contained in the interim and final reports of the 2009 Bushfires Royal Commission, what is alarming is that so little appears to have changed in terms of planning, co-ordination and response strategies in the intervening 70 years despite the enormous scientific and technological advances that have taken place. For example, in the words of one of the survivors of February 2009: 'For ten hours we had learned nothing whatever about a monster fire a few kilometres away' (Manne 2009, p. 24). And this was a fire of such ferocity that certain individuals within key organisations had been predicting and warning about it – but had been ignored – several days and hours earlier (Burns and Eltham 2010; Kissane 2010). None of this would come as any surprise to Perrow (2007) who has written extensively about the 'inevitable inadequacies' of organisations and their propensity to fail. As well, it became patently clear in 1932 and 1939 in Victoria that the construction of firebreaks and the associated practice of back-burning were totally inadequate for dealing with major fires (Griffiths 2001). Small fires can be contained and controlled; large fires cannot.

As we shall see, in addition to a forensic examination of communication and leadership failures, four issues came under particularly close scrutiny in the Commission's hearings: 1) the so-called 'stay or go' policy which places the onus on individual households to take major responsibility for their fire management and response strategies; 2) the efficacy of fuel-reduction (controlled or prescriptive) burning as a precautionary measure to mitigate damage from bushfires; 3) mandated building standards and codes in bushfire-prone areas; and 4) the regulatory land use planning system.

Recent class actions

As well, at the time of writing (November 2010) three legal cases seeking compensation for bushfire damage as a consequence of 'negligence' were before the courts in two jurisdictions. The first, before the Australian Capital Territory (ACT) Supreme Court, heard evidence on behalf of around 600 people seeking compensation from the ACT and New South Wales (NSW) governments following losses from the January 2003 fires that destroyed 501 homes and resulted in four deaths in the Canberra district. Central to this particular case are allegations that state agencies allowed biomass fuel to build up to dangerous levels. The second case is a class action involving 300 people in Victoria's Supreme Court against the electricity company Powercor. The allegation here is that the Remlaw fire in western Victoria in February 2009 was caused by poor maintenance and monitoring of old power lines by the electricity supply company. The third class action – the largest in Victoria's history – is similar but claims negligence against Singapore Power (SP Aus-Net) on behalf of over 700 litigants who suffered as a consequence of the Kilmore East fire that claimed 119 lives in the same month (Silvester 2010). The company has fought back strongly. In its 153-page defence document tabled in the Victorian Supreme Court in October 2010, it claims that the three culpable agencies responsible for contributing to the deaths are the Country Fire Authority, the Department of Sustainability and Environment, and Victoria Police (Fyfe 2010).

The last two cases dramatically highlight some of the consequences flowing from deregulatory, neo-liberal government decisions made in the 1990s to disaggregate and sell off former integrated state-run energy agencies.

Hypercomplex crises

All cases touched upon above are representative of the so-called 'new world' of crises and crisis management which – since 9/11, Hurricane Katrina, the SARS (severe acute respiratory syndrome) epidemic and the global financial crisis, to mention but four recent examples – is now a rapidly expanding sub-field of policy studies (Boin 2009). Indeed, 2010 saw the launch of a new journal – *Risk, Hazards & Crisis in Public Policy* – devoted specifically to this area.

Often formerly labelled 'wicked' problems (Commonwealth of Australia 2007; Brown 2010), relevant topics of central concern to researchers, such as mega-fires, are now commonly referred to as 'transboundary' or 'hypercomplex' crises that, for many analysts, demand a 'radical shift in paradigm and practice' (Lagadec 2009, p. 473). By their very nature, hypercomplex crises cut across multiple policy and institutional settings, and operate at many different scales.

There is no doubting the enormous economic costs of mega-catastrophes. Each year in its *Sigma* series, the reinsurance company Swiss Re, for example, catalogues the damage caused by natural and man-made disasters. At $US26 billion, a recent report ranked 2009 as the eleventh highest year for insured losses since 1970, admittedly well down on the 2005 figure of $US117 billion, the year that Hurricane Katrina struck New Orleans (Swiss Re 2010). In Australia, the COAG (2002) *Natural Disasters* report revealed that, on average, weather-related natural disasters cause over $1.14 billion damage each year to infrastructure, dwellings and livelihoods. The 2009 bushfire alone in Victoria is estimated to have caused $4 billion in damages, and involved a $A1billion insurance payout. In Victoria, the Fire Services Levy is a voluntary commitment. This resulted in many householders and businesses being underinsured in 2009. In Recommendation 64, the Royal Commission urged that this system be replaced with a fairer, property-based levy, as currently operates in Western Australia and South Australia. The Premier of Victoria has subsequently agreed to this change.

We now turn to a closer examination of the February 2009 ('Black Saturday') fire in Victoria, widely portrayed as Australia's worst peacetime disaster. Even though our focus here is mainly on this most recent hazard event, utilising a 'policy learning' framework (Hertin, Turpenny & Jordan 2009), we also make reference to earlier, serious bushfires in Victoria and elsewhere in densely-settled Australia. We revisit the recommendations and policy commitments that were made at the time to ameliorate or prevent loss of life and property in inevitable future mega-fires such as that of 2009.

The February 2009 fires

The seventh of February 2009 ('Black Saturday') in Victoria was described, by those presiding over the Royal Commission, as a day when 'weather conditions were unprecedented in terms of high temperatures, low humidity and wind speeds, following years of drought' (Teague, McLeod & Pascoe 2009, p. vi). In reality, the weather conditions were almost identical to those experienced many decades earlier, in January, 1939, when Melbourne's temperature reached 44 degrees Celsius, or, indeed, in February 1851 when the temperature in central Victoria reached 47 degrees Celsius.

In the days preceding the 2009 fires - the worst in Australia's history in terms of the human death toll (173 people) - a record was set when Melbourne experienced three consecutive days above 43 degrees Celsius (Franklin 2009). As well, January had been the second-driest for that month since measurements started. Such severe drought conditions accelerated the drying out of usually moist mountain ash forests, so much so that fuel loads in excess of twenty tonnes per hectare had accumulated in the forests to the north and east of Melbourne. Ringing metropolitan Melbourne, and excluding closed catchments and national and state parks, these forests have been progressively cleared and fragmented for roads, farmland, townships and scattered residences over time. The process of fragmentation accelerates the drying out of the forests. Nevertheless, the pockets that remain are still a rich source of biodiversity and, in the form of larger blocks, afford invaluable protection to the city's water catchments.

On 7 February, the highest temperature ever recorded in Melbourne - 46.4 degrees Celsius - was reached and hot, northerly wind gusts in excess of 90 kilometres an hour were common across the state. A widely-used measure in fire-fighting agencies is the so-called Fire Danger Index. An index of 50 represents an 'extreme' fire danger. In parts of the state, the index was in excess of 170 (Kissane 2010). Some 300 fires occurred at different times on that day, involving over 12,000 fire-fighting personnel. The majority of these were volunteers in the Country Fire Authority. Many of the smaller fires were successfully extinguished, including one in the heavily populated Dandenong Ranges on Melbourne's eastern fringe that started in the afternoon and had the potential to cause enormous loss of life and property. Serious fires had impacted these ranges in earlier years, most notably in 1962, 1968, 1983 and 1997, and there is little doubt that important lessons had been learned from these events.

In particular, through the 1970s and 1980s the Victorian Government committed around $25 million to a policy of both consolidating and buying back numerous small, inappropriate subdivisions that had been allowed to proliferate in this region of high fire danger. This is a policy, I would argue, that urgently needs reviving in the rapidly-populating peri-urban areas to Melbourne's north and north-east that were ravaged by the 2009 fires.

Government responses

Writing in 2006, Ingram *et al.* noted that

> *Following disasters, governments often clamour to quickly reduce risk, rebuild communities and restore permanence. The pressure to urgently address complex, difficult decisions can result in reactive policies that may increase long-term vulnerability of affected populations.*

Notwithstanding the fact that all but 60 of the 173 people who died in the fires on 'Black Saturday' died in their homes, the strategy to perpetuate the settlement *status quo* by committing to the wholesale reconstruction of damaged or lost dwellings and townships *in situ* was promoted strongly right from the outset by both the Victorian and Commonwealth Governments. Within days of the fires, it became quite clear that a radical rethink of settlement and planning policies was not on the agenda. The then Prime Minister, Kevin Rudd (2009, p. 1), announced that

> *The Commonwealth and Victorian Governments will fund and deliver the site clean-up and demolition program crucial to rebuilding communities devastated by the bushfires…The [Governments] have promised to* ***rebuild fire-affected communities brick by brick*** *(emphasis added).*

Both the then Prime Minister and the then Victorian Premier, John Brumby, were openly in favour of a greatly increased population for Australia. Recent population projections point to cities like Melbourne and Sydney growing by as much as 74 per cent, from around 4 million people today to some 7 million by 2050.

As noted, largely unregulated land use policies currently in place mean that much of the recent – and likely future – population expansion has occurred in high fire-prone areas on the fringes of the major cities such as Melbourne. The outer peri-urban region around Melbourne, for example, has been growing at a rate of 1.8 per cent annually in recent years, over twice the rate of the rest of regional Victoria.

Nevertheless, it is worth highlighting that, by January 2010, only 60 per cent of the households that had lost their homes in the 2009 fires had definitely decided to rebuild on the same site. A further 20 per cent had left the area for good, and twenty per cent were still undecided (Le Grand 2010). In short, a substantial proportion of affected households did not share the enthusiasm of the Prime Minister and Premier of Victoria to rebuild 'brick by brick' in the fire zone. In part, these outcomes reflect substantially higher construction costs following stricter new building codes in high fire-risk areas as well as problems relating to insurance. Around thirteen per cent of all destroyed dwellings, for example, had no insurance cover, and other households have experienced long delays in securing insurance payments.

Also, as detailed earlier, one of the first responses of the State Government was to follow the long-standing tradition of setting up an official inquiry into the fires. The 2009 Royal Commission was established on February 16, with wide-ranging terms of reference. The Premier promised at the time that 'no stone will be left unturned' in the investigation process (Brumby 2009). The Commission's fundamental concern was the protection of human life and it focused in detail on twelve of the most serious fires, four of which swept through relatively densely settled, peri-urban districts to the north and east of metropolitan Melbourne. These were the Narre Warren, Kilmore East, Murrindindi and Bunyip fires. The last three were not fully under control until 10 to 15 March. In total, as noted, there were 173 deaths and 400 people injured (an additional 374 from heat stress) and 2,100 homes destroyed as a result of the fires. Seven thousand people were immediately made homeless. The fires ravaged approximately 430,000 hectares and affected around 78 different communities, in approximately one-third of the state's municipalities (21 in total). In the Kilmore East fire alone, there were 121 fatalities and 1,244 houses lost.

Some 25 per cent of the dwellings destroyed (around 500) were second, or holiday, homes in such tourist townships as Marysville. Unlike principal home-owners, the owners of such properties received no financial assistance from the government (Wilson 2010). For those whose principal place of residence had been destroyed, there was an immediate payment of $50,000. There was also an additional needs-based payment of up to $50,000 for households facing hardship. These sums were drawn from a substantial emergency fund that combined approximately $400 million from individual and corporate donations, a similar amount from the Commonwealth Government, and over $100 million from the Victorian Government. Interestingly, in the disbursement process, no distinction was made between insured and uninsured properties.

Controversially, too, well before the Royal Commission had produced any recommendations, the State Government made a snap decision and acted quickly to relax the existing regulations relating to native vegetation clearance controls.

Under the so-called '10/30 right', introduced as an amendment to the Victorian Planning Provisions, householders in rural and urban fringe municipalities were immediately granted the 'deforestation right' to remove shrubs from within 30 metres of their homes and trees within 10 metres, without having to secure a permit. In addition, with the consent of neighbours, all vegetation could be cleared within four metres of fence lines and local councils could remove roadside vegetation, again without having to apply for permission. Opposition to the 10/30 rule has come from some environmental groups and urban fringe councils on the grounds that such a blanket regulation opens the door for large-scale, indiscriminate land clearing, often for the sole purpose of land development. Nillumbik Council, on Melbourne's northern fringe, which includes both urban and rural fringe areas, for example, recently challenged the ruling in the Victorian Civil and Administrative Tribunal but lost (Bachelard 2010b).

Timing

The *timing* of a disaster and the early recovery effort is always crucial in terms of understanding the local political and response context. For example, the 2009 fires erupted on a Saturday (February 7) in the summer vacation period. Many school buildings were destroyed but no students died at school. From the perspective of the State Government, facing an upcoming election in November 2010 and fearful of another similar tragedy before then, it was deemed essential to be seen to be taking some early, decisive action in 2009, well before the onset of the 2009/2010 summer bushfire season.

But as with the 10/30 ruling, such a hurried policy timeframe collided with the much more measured pace of the parallel, independent Royal Commission that the same State Government had promptly established in February to investigate all aspects of the fires. Even so, mindful of an upcoming fire season, as noted, the Royal Commission produced a set of interim recommendations in August 2009. Almost a year later, and on the eve of the launch of the final Commission report, Victoria's then Opposition Leader, Ted Baillieu, released a list of the recommendations that had not yet been implemented by the Government. This documentation also included some 25 recommendations that had not been acted upon from previous inquiries. Lack of action on the construction of community refuges was particularly singled out for criticism (Austin 2010b; Victorian Liberal Nationals Coalition 2010).

During the course of the Commission hearings, too, it became patently clear that the Victorian Government was desperate to avoid any apportionment of blame against specific individuals in high-level positions of responsibility or public agencies associated with the 2009 fires for fear that this would trigger a political backlash against the incumbent government in the lead-up to the 2010 election. Tellingly, as noted earlier, two senior officials, who had come under intense scrutiny by the Commission and the media about alleged executive failings on the day of the fires, resigned in 2010 prior to the release of the Commission's final report. These were the heads of the Country Fire Authority and Victoria Police.

One of the most controversial issues dealt with by the Royal Commission was of fuel-reduction burning and its role in damage mitigation in major fires.

Fuel-reduction burning

Some fire and forest scientists have long regarded excessive fuel build-up on public land and along roadside edges as a key element in the production and spread of mega-fires. In Victoria, the Stretton Group of concerned forest scientists, for example, has for some years been lobbying for a greatly increased emphasis on fuel-reduction (or prescribed) burning of native vegetation in the non-fire season (Attiwill 2010). To a certain extent the Royal Commission took these arguments into account when it recommended that a minimum target of five per cent of public land should be burned each year across the state, up from the current 1.7 per cent. This would increase the area burned from 130,000 hectares annually to 385,000 hectares.

There were two reactions to this. The first, from the Stretton Group, was that the new target has been set far too low and should be increased to ten or twelve per cent. The second response from such bodies as the Victorian National Parks Association (VNPA) was that our knowledge of fire's place in biodiversity preservation is still extremely limited, especially in relation to native fauna (Gray 2010). Ingamells (2010, p. 10) has argued that 'statewide hectare targets, viewed favourably by the public ... are seriously flawed' and inevitably will lead to a policy largely centred around prescribed burning in remote parts of the state rather than close to settlements. The VNPA favours a range of targets for different ecosystems.

The Victorian Greens have expressed strong opposition to fuel-reduction burning, as have those concerned with the effects of associated air pollution on human health and crop production, especially grapes. Fearful of a major surge to the Greens in the November state election, the Brumby Government deferred its decision on the fuel-reduction recommendation (Lunn 2010).

Challenges and (missed) opportunities

We began by drawing attention to a number of reforms through state intervention that had been introduced around the world over time with the aim of saving lives and making citizens more secure. We shall never know for certain how many of the 173 lives, or the assets lost in the 'Black Saturday' fires in Victoria in February 2009, could have been saved by much stronger state and local government action on many fronts in the weeks and years leading up to this catastrophic event, but in its 67 recommendations the Royal Commission drew attention to numerous public policy failings that need addressing urgently. One of the most significant of these - highlighted in the opening quotation of this chapter - relates to the continued build-up of population in high-risk areas, an issue that could be addressed readily by buy-back schemes and stricter enforcement of land use planning regulations by state and local governments in high-risk areas. This recommendation also featured in the 2004 COAG report (Ellis, Kanowski & Whelan 2004).

In 1996, CSIRO's climate impact group calculated that the 'doubling time' for the number of bushfire days annually was 70 years into the future. Five years later they revised that figure down to 30 years (Langsam 2010). As a consequence of climate change, there will be more mega-fires in the future and inevitably they will be of such magnitude that it is futile to attempt to fight them; but surprisingly the Royal Commission made no recommendations relating to greenhouse gas emissions. As is already done routinely for zones vulnerable to flooding, the immediate challenge is to nominate areas that are deemed 'unfit for habitation' as a way of reducing the future death toll (Buxton 2010; Stewart 2010). But identifying such areas and foreclosing the settlement option clashes head-on with the strong desire of local government authorities to expand their revenue base and with what many in Australia believe is their long-established 'right' to live wherever they choose, adopt the long-established 'stay or go' tactic in the event of a fire, and be left alone to 'tame' nature. This applies particularly to people wishing to make the 'tree change' transition. In a radio interview, the then Premier of Victoria, John Brumby, commented that '… I don't want to be in the business of telling country towns and country communities that we're going to close you up' (quoted in Wilson and Rintoul 2010, p. 4). The recommendation to place more emphasis on the construction of community refuges also accords with this view.

As a conservative estimate, fulfilling the Royal Commission's recommendations has been put at A$35 billion. This would appear to be a small price to pay for saving untold lives in the future, and indeed the Commissioners have argued that their costings are responsible. But the battle-lines are already drawn in terms of whether this is a reasonable investment and indeed who should pay. To take one final example, overhead power lines came in for much criticism during the Commission's hearings, but SP AusNet revealed that they would face a bill of $7.5 billion to replace the overhead wires in their district. This, in turn, has prompted some commentators to argue the multiple benefits of decentralised electricity generation from sources other than coal (Miller 2010). As is so often the case, crises also provide potential new opportunities.

Chapter 11

Boundless plains to share: Why no population policy?

MARK O'CONNOR

Introduction

It is a myth that Australia is a big country with 'boundless plains to share'. The human and environmental consequences of a population growing over the past 50 years at an average of 1.6 per cent a year are serious. Such a rate is about four times the current OECD average and is high even for Third World countries (compare Indonesia's 1.2 per cent in 2008). Yet powerful business forces seek to increase this rate. At the 2 per cent growth rate experienced in 2008-2009, Australia's population would double roughly every 35 years. By 2047 there would then be some 17 million more Australians than the 28.5 million predicted in the *Intergenerational Report* (Australian Government 2007) on which much planning was based, and an unfeedable100 million before 2100. Rapid population growth leaves Australia little chance of reducing its greenhouse emissions or its rates of environmental damage (Diesendorf 2009). Traditional conservation strategies remain vital, but the issue of human over-population must now be at their forefront (McKee 2009). Australia's missing population policy is perhaps its most serious neglected environmental issue.

Is Australia a big country?

> *Are Australia's natural resource endowments, including water, capable of sustaining a population of 35 million? . . . Must it mean an even greater loss of biodiversity – difficult as that might be to imagine, given our history of species extermination? (Henry 2009).*

Though the semi-desert nature of most of Australia has long been known, the nineteenth century saw the emergence of influential groups of growth advocates – then known as boomers or boosters – who argued that Australia's large area meant it must soon reach the population of another Europe, or a new United States of America (US). They distributed maps of Australia with outlines of European countries super-imposed, showing for instance that most of Western Europe could fit into Western Australia. In the 1920s the geographer Griffith Taylor countered by producing maps with Australia turned upside down and superimposed on Saharan and sub-Saharan Africa – a better fit in area, and in rainfall and soil fertility. He also predicted, correctly, that Australia's population in 2000 would be close to 20 million, and within the existing 'boundaries of settlement' (Powell 1988, p. 131; Watson 2008).

Inland Australia feels spacious. Yet an over-populated country is not necessarily a crowded one. Even China has empty deserts. Over-population is best defined as a condition in which the community cannot live on its environmental 'interest' and so must deplete its 'capital'.

In human terms, Australia is not so much a big country as 'a small country with big distances' between the productive areas (Seddon 1997, pp. 222-223). Australia's agricultural frontier closed in the 1890s – and even then the land was not truly empty but seized from Aboriginal owners. New arrivals since then have had to pay through the nose for land, even for housing blocks in the cities. Yet a popular myth that Australia is still in its pioneering adolescence leads many Australians to accept constant population growth as normal (O'Connor 2008).

Even Australia's agricultural areas are not so large, or fertile, as proponents of 'big Australia' claim. Australia suffers a capricious climate with fierce alternations of drought and flood. It is the world's flattest, driest, least fertile and most ruthlessly cleared habitable continent. Only ten per cent is arable. In a good year it grows less wheat than France, in a bad year less wheat than Britain (Flannery 1994b).

Australia at present produces nearly three times as much food as it consumes, and most Australians assume they will remain rich in food. Yet a mere doubling of population, which at 1.8 per cent annual growth would occur in some 39 years, would halve Australia's per capita resources of land, fresh water and seafoods, and thus cut its food exports drastically. At the same time, production is likely to decrease as Australia's farmland becomes eroded, salinised and acidified, and as the peak in the availability of oil drives up the cost of imported fertilisers. Every tonne of wheat still costs some fifty tonnes of eroded soil. Because nitrate fertilisers are made directly from oil and natural gas, their cost may fluctuate with the price of oil, which rose from $US10 a barrel in 1999 to over $100 before it was checked by the global financial crisis of 2008 (Lines & O'Connor 2010 p. 127). Australia's oil will be gone by around 2020 (Australian Government 2010, pp. 117, 138). Phosphate fertilisers are equally critical, and increasingly scarce. The processing and transport of crops also depend on fossil fuels.

Tim Flannery has offered a rough calculation of Australia's long-term carrying capacity. He suggested that, with currently foreseeable technology and with our present habits of consumption, we might in the long run safely support between 8 and 12 million people (Flannery 1994a). If so, this will be bad news not just for Australia but for some 80 countries that cannot feed themselves and are dependent on the few countries that have a surplus. (Flannery did not factor in the rate at which some other countries are now buying up farmland in Australia to solve their own food problems). His estimate should be compared with an actual Australian population of 22.5 million in early 2011, and a projected 35.9 million as early as 2050, even if net annual migration can be kept to 180,000 (Australian Government 2010, p. 8).

The 'empty country' myth is often mixed with a moral argument that Australia would be 'selfish' to seek to be less over-populated than other countries. This has no basis in international law, which assumes nation states own their own countries. Indeed most of the countries at the top of the world's per capita wealth table have populations under ten or twenty million, like Norway and Sweden, and are admired rather than blamed for not letting their populations outrun their resources (McNamara 2009). By contrast, Hitler's claim in the 1930s that the expanding German population should claim living space or *lebensraum* from less dynamic countries was widely condemned, as was Japan's expansionism at that time. A vaguer version of this argument, still human-centred, holds that since 'population is a global problem', it is contrary to 'global equity' for any country to solve it on its own. Yet there is no competent global authority, and the ecosystems threatened are mostly not global but local. In fact, the world's nations are more like a free enterprise system, where those who act prudently can benefit from their choices while helping and setting an example to encourage others (Lines & O'Connor 2010, pp. 56-57, 61-63 & 70-71; Carr 2010 pp. 3, 31). In contrast 'population socialism', if widely accepted, might require all nations to 'sign on for the population explosion' (Carr 2010), and cause disaster not just for other species but for humans.

After Australia separated from Britain in 1901, many Australians still felt an obligation to take in United Kingdom (UK) migrants, granted that UK families had contributed men, money and lives to the fleets and armies that had long protected Australia. Later, after WWII, there was a strong sense of obligation to take in people from a shattered Europe, and also a military argument that Australia must 'populate or perish'. Of course such arguments are human-centred and ignore other species.

A newer version of the moral argument claims that Australia must take in millions of people of non-European races to make up for those it did not accept during the years of the White Australia policy (1901-1973). While the White Australia policy was certainly reprehensible, non-refugee immigration to any country is a privilege and not a right. In any case Australia has long ceased to discriminate by race in its selection of immigrants, and has taken in generous numbers from surrounding countries. An odder moral argument, sometimes heard from the Green left, is that Australia, having displaced the Aborigines through immigration from Europe and reduced them to a minority in their own country, must now make amends for such racism by taking in comparable numbers from the other continents. Aborigines often resent this argument, and complain that their protests against immigration are ignored (O'Connor 2008; Lines & O'Connor 2010, pp. 176-179).

Idealists sometimes portray Australia as a large rich country that should share some of its wealth with 'poor' countries such as India, China or Indonesia. In reality these countries are not at all poor in natural resources, but much richer than Australia. That is why they can hold such great populations. Virtually none of Australia's farmland compares with Java's lush double-monsoon regions with metres of rich volcanic soil. Even in economic terms, Australia's richness is something of a myth. India, for instance, according to *Forbes* Magazine, has as many middle class people - some 300 million - as the total population of the US (David 2007). It also has more than 100,000 persons who are US-dollar millionaires. Some people imagine that Australia could solve the rest of the world's environmental problems by taking more immigrants. This is not so. India and China together grow each year by about the total population of Australia.

The over-population of Asia is also quite recent. Thus Java, which held 124 million humans in 2005, held less than 5 million in 1800. This growth was produced, in two centuries, by an average annual increase less than Australia's recent rates (Lines & O'Connor 2010, p. 29). The human history of even the most crowded countries starts with a small population and usually with an intact environment and abundant natural resources. Yet once countries pass 20 million, as Australia did in 2003, they are out of the tiddlers' pool - and only two or three doublings away from becoming another obese behemoth of around 100 million.

Australia's population boosters remain active. Today most pay only lip-service to the myth of an empty continent awaiting farmers. The big money is to be made in housing and in land speculation, as well as from retail sales and construction. Newspapers are fed repeated stories about how we are desperately short of skilled and willing workers - alternating with pieces on how we are desperately short of major projects to provide employment. The proposed solution is a spiral of increasing population and increasing construction that the United Nations' (UN) former chief demographer Joseph Chamie has described as 'Ponzi demography' (Chamie 2010). Murdoch's *Australian* newspaper aggressively demands high immigration and rapid population growth. Yet, as several enquiries have found, 'the economic case for rapid population growth through immigration is surprisingly weak' (Gittins 2010; Carr 2010, pp. 21 & 34ff). Related claims that we must have high migration to provide labour, or skilled labour, or to stop the population ageing are also dubious (Carr 2010, pp. 22-24, 25-26; Productivity Commission 2010; Gittins 2010).

Extreme population growth?

> *The quality of all aspects of our children's lives will be maximized if the population of Australia by the mid-21st century is kept to the low, stable end of the achievable range, i.e. to approximately 23 million (Australian Academy of Science 1994).*

Australians know their country has a growing population; few realise how unusual its growth-rate is. Most other OECD countries have slowed their population growth, and some have begun to reduce their populations. By 1997, 29

European countries had population growth rates around zero, the actual rates ranging between -0.6 per cent (Russia) and +0.3 per cent (for example, France and Norway). Japan's recorded population change was negative for the first time in 2005.

Australia averaged over two per cent annual population growth during the 1950s and 1960s; well over one per cent through the 1970s, 1980s and 1990s; and still over one per cent and rising through the 2000s. By early 2007, late in the Howard Government's tenure, this had reached a steep 1.5% – well above the Asian and world averages, which are both around 1.1% (CIA 2011). By mid-2008 Australia reached 1.6%, and in early 2009 it passed 2%. However, by mid 2011 Australian Bureau of Statistics (ABS) figures suggested it was back to about 1.6%.

Both net migration (the surplus of immigrants over emigrants) and natural increase (the surplus of births over deaths) had been driving this increase, each adding about 125,000 persons a year. Yet by 2009 annual net migration (which included many 'temporary' workers who had stayed, plus a large influx of overseas students attracted by the belief that they would receive permanent residency) reached 300,000. In three years net overseas migration (NOM) 'more than doubled, from 146,800 persons in 2005-06 to a preliminary NOM estimate of 298,900 persons in 2008-09, the highest on record' (ABS 2010). However, a tightening of permanent residency for students led to a reduction in population growth in the third quarter of 2009 from around 2.1 per cent a year to 1.8 per cent. At this rate, if maintained, Australia would still pass 100 million before 2100 (Lines & O'Connor 2010, p. 135).

Does population growth damage environments?

> *It is self-evident that the massive growth in the human population through the C20th has had more impact on biodiversity than any other single factor' (UK Chief Scientist Sir David King 2006).*

Human beings now compete with most other species for space and resources, and there is an obvious chain of connections. Population growth drives clearance of native vegetation (the prime cause of biodiversity loss) in two principal ways. It does so *directly*, through urban expansion to create more roads, houses, offices, hospitals, shopping malls, and so on (often on the richest land or the coastal fringe); it also does so *indirectly*, through the expansion of industries (including agriculture and timber extraction) that generate either consumables for Australia's human population or export earnings to offset the growing imports consumed by the same population. As domestic population grows, so does domestic demand for food and fibre, which reduces the quantities available for export. This adds pressure to clear more land.

Empirical studies bear out these connections. Globally, 'Human population density, and agricultural land use are the best combined predictors of threats to species of mammals and birds. [Hence] ... Without addressing the human overpopulation issue, all other measures will eventually be for naught' (McKee 2009, p. 308). 'The key variables which influence environmental quality are population numbers, human behavior, and the characteristics of the environment itself' (Burnley, Fagan & Murphy 1997, p. 98).

Australia's first independent *State of the Environment* report, published in 1996, insisted that it is essential to curb population growth. It itemised the nation's ecological disasters and noted:

> *The overwhelming causes of the decline in Australia's biodiversity result from the human population, their lifestyles, technologies and demands on natural resources. The situation continues to deteriorate as population and demands on natural resources increase (Australian State of the Environment Advisory Council 1996).*

The two subsequent reports paint the same picture: worsening environmental conditions, particularly in regard to water shortages and quality; biodiversity loss; and soil degradation. Except for urban air quality, all the indicators point downwards (ASEC 2001, 2006). The Australian Government's official report to the UN has identified 'population growth and unsustainable development' as the major current and long-term threats to Australia's biodiversity (Australian Government 2009).

Yet faced with politicians who are under party orders, as well as business and media pressure, to deny the link between population growth and environmental damage, appeals to logic or common sense are not enough. Some obfuscate, claiming that 'the issue is not population but consumption'. This implies that Australians will soon compensate for their increased population by a corresponding reduction in most forms of per capita consumption (and thus of per capita demands on the environment). But there is at present neither a realistic likelihood nor a political intention of reducing per capita consumption. Labor and Liberal governments alike compete to assure the public that their economic skills will deliver ever higher affluence. Both residents and new arrivals at present seek predominantly to improve their lifestyle by increasing personal consumption. This is encouraged by massive expenditure on advertising.

Australians appear to be the world's worst greenhouse polluters, producing about 26 tonnes of CO_2 per person per year. In contrast, Europe produces ten tonnes, China four tonnes, and India two (UNEP 2007, p. 8; Lines & O'Connor 2010, p. 20).

Voluntary reductions in individual consumption tend to have little overall effect, because measures to prevent or forestall environmental problems pursue *common goods,* and so require the participation of all (or most). Public frugality (which includes policies to set limits to population growth) must precede, not follow, private thrift (Daly 2002). Otherwise, private persons who practice thrift may merely subsidise and prolong public profligacy. Similarly, more efficient technology can slow the damage for a while but is often subject to diminishing returns. The first ten per cent of savings come easily, but each subsequent ten per cent is harder, especially if it involves replacing infrastructure.

> *If the population is allowed to expand on the back of efficiency gains in resource use, this* ***decreases reserve capacity and resilience****. The population will then be far more vulnerable to shortage shocks in its resource base; it will be in greater danger of collapse.' (Lardelli 2009, p. 2; cf. Carr 2010, p. 14).*

Australia has little hope of meeting the Rudd-Gillard government's emissions target – a 60 per cent cut by 2050 – unless it halts its predicted population growth (of about ten million) over the next four decades. Even if Australia's population only climbed to 31.6 million by 2050, this would result in 276 million tonnes of greenhouse pollution more than if population stabilised at around 22 million by 2050 (Birrell & Healy 2008).

In 2010, Australia's largest conservation body, the Australian Conservation Foundation (ACF), formally applied to the minister for the environment to have Australia's population growth classified as a threatening process for the environment (Berger 2010).

Surprisingly, those who refuse to accept that population can be a problem include some environmentalists. Population blindness is widespread (Diesendorf 2009).

The human costs of population growth

> *I'd like to see a situation where you can get a seat on a bus (Tony Burke, Australia's first Minister for Sustainable Population, on Sky News, 4 July 2010).*

Environmental scientists might define a sustainable human population as one that (a) can supply its own wants in perpetuity from its own natural resources, and (b) no longer brings ongoing damage to natural environments. Politicians have much looser, even trivial, definitions, and an extremely short-term focus. ('Sustainable' may simply mean what they can get away with till the next election.) For them, a major issue is the infrastructure debate.

This debate requires us to distinguish two sets of population problems. One set is caused by a country's absolute population size. (Most problems for other species are of this kind.) The other set is caused by the *rate of population growth*. Thus Jane O'Sullivan suggests that the extra infrastructure cost for each person added to the population is probably around US$200,000; although William Bourke has argued that it may be closer to $A500,000 (O'Sullivan 2010; Bourke 2010). Population growth rates of more than two per cent a year require more spending on infrastructure (notably trains, buses, bridges, roads, drains, sewers, water mains, electricity grids and hospitals) than governments dare ask from the taxpayer (O'Sullivan 2010). Hence infrastructure and services are overstretched, and budgets (including for conservation) are often slashed.

As well, population growth adds to many urban problems: pollution, congestion, commuting times, high house prices (Burnley, Fagan & Murphy 1997, p. 89) and the resulting need to work long hours to pay off crippling mortgages, expensive and inconvenient rebuilding of cities, and loss of parklands. It also adds to the future dilemma of how to supply Australians with oil, electricity, water and food while at the same time trying to reduce emissions and preserve environments (Foran & Poldy 2002a & 2002b; Carr 2010, p. 31).

Such problems tend to compound. Lardelli (2009) comments:

> *As Melbourne's population grows it sends out pipelines like invasive roots to suck water out of what used to be distant river systems – but those river systems, such as the Goulburn River, are already being used to grow the food that Melbourne needs . . . Population growth lies at the core of all questions of sustainability. It is growth of the human population and the increasing resources that it requires that drives the destruction of habitats, the increasing levels of pollution (especially CO2), and the accelerating depletion of finite resources. It is impossible to stop exacerbating these problems if we do not stop population growth.*

Australia is one of the world's most urbanised countries, with 90 per cent of the population living on two per cent of land. Hence the effects of population growth on human beings are felt intensely in its cities. For instance, in 2006 *The Sydney Morning Herald* reported that Sydney's councils had abruptly been instructed to accommodate an extra 1.1 million people within 25 years (SMH 2006). The councils protested in vain that this would destroy their plans for maintaining Sydney's once-impressive environmental amenities. Yet large increases in Australia's and Sydney's projected population since 2006 mean that even these quotas will prove inadequate. Developers often claim that the result will be 'vibrancy' as opposed to 'stagnation'; but many residents see, or fear, a continually densifying and degrading cityscape.

There are also psychological costs. The Labor MP Kelvin Thomson claims:

> *Our cities are too large. They dwarf people. The sheer scale of them is overwhelming for some, who lose the plot and fall victim to mental illness or drug and alcohol abuse. And for the rest of us the madding crowd swells every year, giving us that little bit less room ... In my home city of Melbourne, a lot of people of goodwill have supported high rise as preferable to urban sprawl. What they don't realize is that it isn't halting any urban sprawl at all. Suburbs continue to march out onto the horizon. Property developers are having their cake and eating it, too (Thomson 2009a, p. 8106).*

There is no central register of such 'diffuse costs' to our quality of life, and no easy way of summing up 'intangibles' (Productivity Commission 2010). Hence environmentalists, focused on wild nature, have sometimes done too little to protect established suburbs from destructive redevelopment. Indeed, they often help complicit state governments by leading debate away from population growth ('not the real issue') towards speculation on whether densification (which may destroy the human scale and amenities of a city) or decentralisation (pushing people out of the city or to its fringes) is 'the solution'. In practice, population growth commonly brings both.

It also greatly increases a city's demands and impact. Denser suburbs may *permit* better public transport, but the assumption of a strong link between density and effective public transport has been seriously challenged (Mees 2009). Yet at present, in practice, they bring a multiplication of cars and emissions. In any case, the energy costs of high-rise living and of rebuilding existing suburbs piecemeal (with great nuisance to residents) may outweigh any energy savings on transport (Myors 2005; Glazebrook, Rickwood & Searle 2007).

Yet most politicians prefer to believe that technology or densification - or even decentralisation - will 'de-link' population from such problems. In December 2010, the Minister for Sustainable Population, Tony Burke, released an *Issues Paper* in which he implied large claims for regionalisation:

> *The key to understanding policy for a sustainable Australia begins with a principle of regional difference. In some communities infrastructure and services have not kept pace with population growth, while in others employers are facing a chronic shortage of skilled labour (Burke 2010, p. 1).*

Yet we are now in the context of the peaking in the availability of oil, which may have occurred in 2008, and is likely to bring at least some years of energy shortage before alternative sources can be exploited (Aleklett 2010; McKibben 2010). There is also an extensive literature about the risk that dense cities may turn into unsustainable 'killing zones' in a world that runs short of energy and hence of food, or the means and will to transport food (*Energy Bulletin* 2010; Goodchild 2010; McNamara 2010-011). In 2010, the Department of Immigration and Citizenship warned the incoming Gillard government that it expected about 90 per cent of immigrants to continue settling in Australia's major cities, and that its medium-term predictions for immigration exceeded those on which Treasury's 2010 *Intergenerational Report* was based (Scott 2011). Yet making Australia's main cities bigger or denser is unlikely to make them more sustainable; decentralising a growing population into regional cities, even if feasible, may over-load the infrastructure of those cities; and moving the population to small towns or farms seems implausible.

Debating population: recent politics

> *Australia's major political parties have had no population policies in recent times, merely adjusting immigration rates periodically (Andrew McNamara, former Queensland Minister for Sustainability and Climate Change, 2009).*

The Whitlam Labor Government (1972-75) reacted to the first global oil shock in 1973 by seeking to limit immigration and population growth. Its attempt to buy back Australia's energy resources for self-sufficiency attracted fierce opposition from business lobbies and media barons. This was the last Labor government to risk seeming anti-business. The incoming conservative government of Malcolm Fraser promoted rapid immigration, ignored energy constraints, and was made to seem prescient by the global easing of the oil shortage.

Spin doctors now represented criticism of immigration levels as an 'insult' to ethnic groups, even though polls repeatedly showed that immigrant Australians also disapproved of current levels of immigration. This tactic was repeated by

Labor, now in favour of high immigration and the off-shore sale of energy assets, when it returned to power in 1983. Immigration levels began to slide steadily upwards. Those who pointed out that Australia (or Australia and Canada) had the world's highest per capita immigration were regularly accused of 'racism' or of 'scapegoating immigrants'. It was not until 2009 that the Federation of Ethnic Community Councils of Australia (FECCA) publicly dissociated itself from one such accusation and said it supported open debate on immigration levels (Lines & O'Connor 2010, pp. 108, 145, 205, 215).

In 1994 the Australian Academy of Science spoke out. It advocated: limiting Australia's population to 23 million; the encouragement of sex education and birth control ('every baby a wanted baby'); and holding net migration in the 'responsible range'. This meant below 50,000 a year (about a half of what it was during most of the Hawke-Keating Labor period, about a quarter of the 190,000 to which the Howard Government would push it by 2007, and about a sixth of what it reached during the Rudd Labor Government). Aware that spin doctors had encouraged the public to confuse immigrants and refugees, the Academy pointed out that a much expanded refugee program could fit within the 50,000 cap (Australian Academy of Science 1994).

In the same year, a committee of Federal Parliament published a detailed investigation of the population issue (HRSCLTS 1994). This stopped a little short of endorsing a 23 million maximum, but recommended a formal population policy be created. Its chair Barry Jones, the Australian Labor Party's (ALP) federal president, told Parliament: 'Immigration intake must be an instrument of population policy. Population policy cannot be a long-term unconsidered side effect of *ad hoc* immigration practice' (Jones 1994).

Yet the Keating Labor Government did nothing; and the Howard Government later responded that no formal population policy was needed, because its existing policies would lead to only some 23.5 million Australians by 2051 (Australian Government 1999). It, thereafter, both introduced the baby bonus and significantly increased immigration.

Howard's immigration minister, Philip Ruddock, later commissioned the CSIRO to report on 'Options to 2050 for Australia's population, technology, resources and environment'. This led to the major report *Future Dilemmas* (Foran & Poldy 2002a) and its companion booklet *Dilemmas Distilled* (Foran & Poldy 2002b). The Howard Government found the report's warnings against continued population growth unwelcome, and attempted to suppress it (Fullerton 2002).

Meanwhile, although births were twice deaths, the government ran a scare campaign about Australia's 'low fertility' and 'falling population'. In 2004, it introduced a $3,000 baby bonus (later increased to $4,000) to persuade couples to have more babies. After Howard's defeat in 2007, the Rudd Labor government increased the bonus to $5,000 to prove it supported 'working families' (Lines & O'Connor 2010, pp. 126, 210).

Until late 2009 the media remained indifferent or confused, while moralising commentators persistently confused immigrants with refugees. (In reality Australia's combined refugee and humanitarian intake of less than 14,000 a year is negligible beside a gross migration intake that has sometimes topped 500,000.) Yet a Monash University survey in December-January 2009-2010 found 72 per cent of respondents believed 'Australia does not need more people' (Betts 2010b, pp. 43-44).

In fact, the ice was melting. New media, such as the website of Sustainable Population Australia and the Can Do Better site, were outflanking (and in some cases converting) the old media (Australian Population Websites 2010). In February 2009, former NSW Premier Bob Carr launched *Overloading Australia: How governments and media dither and deny on population* (Lines & O'Connor 2009). Running through four editions in under two years, the book helped convert some who had wavered on the issue. It recapitulated arguments from many eminent scientists but added refutations of the common evasions and counter-arguments. Scientists, because they respect expertise, too often fall silent when vested interests or evasive politicians move the debate away from their area of expertise with remarks like: 'That's very sad about the environment, but what about the terrible labour shortage, and the cost of an aging population if we cut migration?' (Gittins 2010; Carr 2010).

In late 2009 there was some media interest when advance figures began to leak from the 2010 *Intergenerational Report* (Australian Government 2010). These projected an unexpectedly high population of 35.9 million by 2050. At lower rates of population growth, many people may not notice the deterioration in their lifestyle, but now growth at two per cent a year was bringing deterioration too rapid to ignore – for instance in health services and in commuting times. Voter anger rose sharply, but was not yet clear from the polls because voters did not know which party to blame. Then, in October 2009, Kevin Rudd declared himself a believer in a 'big Australia'. There was an outcry, and his formerly very high personal popularity began a rapid decline. This accelerated when, in April 2010, he was forced to abandon his promised greenhouse gas emissions cuts (Lines & O'Connor 2010, pp. 2-3).

In January 2010, the Australian Broadcasting Corporation's (ABC) *7.30 Report* ran a week-long series of stories on the population debate, temporarily ending ABC TV's quasi-blackout on the issue. In March 2010 the Greens leader, Bob Brown, moved in the Senate for an enquiry into the costs of population growth, and, as mentioned above, the ACF designated population a threatening process. In April, Rudd appointed a Minister for Population, Tony Burke, who sought ways to defuse public concern. Burke claimed that urban voters feared population growth but that rural voters were concerned about *depopulation*. In reality, non-metropolitan Australians are significantly more opposed to a 'big Australia' than urban ones (Betts 2010a).

In 2009, the federal Labor MP Kelvin Thomson, and then in 2010 the philanthropist and former Australian of the Year Dick Smith, began to speak out regularly on population. Smith sent copies of *Overloading Australia* to every Australian state or federal MP and to every mayor, and helped fund a documentary *Dick Smith's Population Puzzle*, which ran on ABC TV.

In May 2010, the Public Affairs Committee of the Anglican Church in Australia stated that it was now a matter of morality for Australians to limit their population so as not to steal from the future (Still 2010). In June 2010, delegates at the Australian Local Government Association's National General Assembly called for a federal population policy. Also in June 2010, Julia Gillard displaced Rudd as prime minister and immediately declared (with an election imminent) that she did not believe in a 'big Australia'.

In late 2010, the Immigration Department received a sequel, which it had commissioned, to the *Future Dilemmas* report. Called *Long-term Physical Implications of Net Overseas Migration* (Sobels *et al.* 2010), it looked systematically at the population sizes that different levels of immigration would bring, and then at the problems for resources, energy, quality of life and environment that would follow. Pro-growth forces attempted to censor it both before and after its completion; and the Department released it on Christmas Eve, which resulted in its warnings receiving little media attention (O'Connor 2011).

In May 2011, Minister Tony Burke released the Gillard government's population policy. This stated that the government rejected 'big Australia'; but the policy was widely derided by critics on both sides of the debate because it offered no precise population figures or targets. Net migration was now around 180,000 a year. This was less than under Kevin Rudd, but was the figure Treasury had assumed when making its original big Australia prediction of 35.9 million by 2050.

An environmentally sound population policy?

> *Any talk of sustainability without a commitment to population stabilisation is not just spin; it is a dangerous lie… [Yet] There is simply no language in our existing political lexicon to explain to voters that we have borrowed more than we can afford from both the past and the future to pay for today's lifestyle (McNamara 2009, p. 26).*

The challenge for environmentalists is to propose a workable policy for capping Australia's population – and work out how to persuade politicians to adopt it. By 2011, politicians knew that most voters opposed the 'big Australia' option, yet public opinion does not necessarily sway politicians. They need the public's votes at election time, but they also need campaign contributions and favourable attention from the media, whose owners favour growth. (Indeed, the Murdoch press aggressively targets critics of growth.) Trapped between business donors who see big dollars in high immigration and an electorate that sees high immigration as the giving away of shares in a country they own, most politicians, once in office, prefer not to see the problem. It soon became clear that neither the Gillard government nor the Liberal-National Coalition opposition intended to cap Australia's population.

Nor, surprisingly, did the Australian Greens. In early 2011, their population policy stated no desirable target for Australia's population, and no maximum figure for net migration (Greens 2011). Many of its clauses offer dubious excuses for not having a clear policy, for example, that 'environmental impact is not determined by population numbers alone'. There is a constant struggle within the Greens between those who put the environment first and those whose passion is, in the policy's words, 'global social justice' (Lardelli 2010; Mcnamara 2010-2011, p. 27).

Carr (2010, p. 29) does not recommend a precise population target, but argues that Australia needs an independent 'sustainability commission'. Population growth could then be made dependent on improvements in sustainability indicators. Yet many environmentalists believe a population policy must set rigorous targets if politicians are to respect it.

Such a policy could promise an end to indefinite population growth, but not an immediate stop. By 2011, two proposals were on display: Kelvin Thomson's 14 point plan for population reform, and that of the Stable Population Party of Australia (SPPA), which formed after the 2010 federal election (Thomson 2009b, SPPA 2010).

Constraints on proposals for population reform

Australia's natural increase in 2011 was still strongly positive, adding over 100,000 persons a year – and no party proposes to restrict the number of children Australian parents may have. Still, natural increase is expected to fall. It is often said that 'replacement fertility' under Australian conditions is 2.05 children per woman – the extra 0.5 allows for deaths in childhood and for a slight excess of males. The figure in September 2011 was just under 1.9 children per woman.

However, we should not fall into the 'replacement rate fallacy' – the myth that a fertility rate of less than two surviving children per woman means that births are (or will soon become) fewer than deaths. This fallacy relies on imagining that children replace their parents. They don't – or we would all be born orphans. It would be more nearly true to say that children replace great-grandparents, which is why it often takes a few generations of so-called 'replacement fertility' before births cease to exceed deaths. Still, predictions are that this may occur in Australia within twenty years. Natural increase, especially if the baby bonus is removed, might add only another one million to Australia's population, which was 22.5 million at the start of 2011 (O'Connor 2009; Productivity Commission 2010).

Net migration is the other, and larger, long-term danger to population stability. Most Australians have small families and do not want a 'big Australia', but in recent years the government has been 'choosing to have less control over the immigration program' (Carr 2010, p. 30). In 2009 and 2010, net migration was adding anything up to 300,000 persons a year, and business groups were pressing for further rises to compensate for expected falls in natural increase. Yet migration is, in theory, much easier for Australian governments to control.

Thomson's '14 Point Plan for Population Reform' of November 2009 is rather moderate. It would abolish the baby bonus, but would allow a net surplus of 70,000 more immigrants a year than emigrants. Within this quota he would actually increase the Refugee and Humanitarian Program from 13,750 to 20,000 per year. Australia's population would climb by some four million, the equivalent of another major capital city, before eventually stabilising at perhaps 26 million – after natural increase becomes negative. The plan is, in Thomson's words, 'an alternative to runaway population' (Thomson 2009b).

The SPPA's policy states that it is committed to a 'stable, sustainable, open and tolerant Australia' (SPPA 2010). However, it offers quicker relief from population growth via a larger cutback in immigration. Its two chief proposals are:

- phase out the baby bonus, and
- run immigration at around 50,000 per annum (which, it notes, is around the average per capita rate for western countries) and maintain Australia's current annual refugee intake of 13,750.

This quota of 50,000 migration places per annum is gross, not net. Like the original policy of the Australian Democrats, it would produce roughly the same number of immigrants as emigrants. The aim is 'to stabilize Australia's population at around 23-26 million through to 2050'. The rather large error margin '23-26' underlines that the policy's effectiveness depends somewhat on future fertility rates.

Indeed, both policies depend for their success on natural increase (births minus deaths) falling; and Thomson's policy relies on it becoming negative to balance a continuing net inflow of immigrants. If Australian couples – or as demographers prefer to say, Australian women – increase their fertility towards or past the 'replacement rate', then these policies will work more slowly or not at all. Conversely, if falling fertility by 2050 produces a large excess of deaths over births, net migration could be raised. Thus such policies have an implied expiry date. SPPA's phrase 'through to 2050', recognises that it cannot foresee the circumstances of the indefinite future.

Note that neither scheme presents its target, whether 26 or 23-26 million, as Australia's ideal population. Neither has criticised the Academy of Science's 1994 finding that 23 million is the highest safe figure for Australia. Rather, they seek a target that is still demographically and politically possible. In fact, to specify a precise optimum population size for Australia would be difficult, since new technologies or changing circumstances (for example global warming, social breakdown) might change the calculations. Thomson and SPPA feel it may be simpler in practice to reach agreement that we have passed the optimum – that the boat has sailed downstream too close to the rapids and should moor at the first suitable landing.

Chapter 12

Australian ENGOs and government: Strengthening democratic advocacy

JOAN STAPLES

Introduction

In mid-1987, there were rumours flying that the Australian Labor Government was about to go to an election. At this time, I was stopped in the corridors of Parliament House by the 'numbers man' to Prime Minister Bob Hawke, Senator Graham Richardson, who asked: 'What is the big one we have to do for this election? Is it the wet tropics?' As environmental lobbyist for the Australian Conservation Foundation (ACF), I simply replied: 'Yes'. Within a few days, the election was announced and the Prime Minister publicly undertook that his government, if returned, would stop the logging that was destroying Queensland's rainforests by nominating them for World Heritage listing. This was the only lobbying I ever needed to do with Senator Richardson on the wet tropics. However, I had been working with him over a period of eighteen months on other major national issues and by this time, I was confident that the Australian Labor Party would promote the environment, and specifically four major issues, including the wet tropics, as a major plank in their 1987 election platform. As a result, the ACF and The Wilderness Society (TWS), the two major national environmental non-government organisations (ENGOs), endorsed and encouraged a vote for Labor at the election.

Eighteen years later, in 2005 under the Howard Government, the ACF and TWS no longer talked to government ministers, nor did they meet with any government representatives (Tupper 2008). Both felt under siege from attacks in parliament on non-government organisations (NGOs) generally and on ENGOs in particular. The then Director of TWS describes his antagonistic relationship with the government as having 'daggers drawn' (Marr 2008). These two situations, nearly two decades apart, represent very different relationships between ENGOs and the Federal Government, yet neither is ideal for democracy or to sustain the sector through changes of government.

There has been a failure, it would seem, by ENGOs to properly recognise where they need to position themselves in the Australian polity and, as part of this failure, the history of endorsement of Labor under Hawke and repression experienced under Howard continues to haunt them. A fresh analysis of their position in relation to the government is required in order to revitalise their advocacy, and it is suggested here that ENGOs need to ask themselves what sort of relationship with government will strengthen democracy and the important advocacy role that the ENGO sector can play, but also be sustainable as governments come and go.

A lack of clarity about its democratic role comes at a time when there are urgent policy issues requiring attention, not the least being climate change, at a time when the sector has appeared to be in some disarray. The key national organisation, ACF, split from most of the sector in May 2009 when its Director, Don Henry, supported the Rudd Government's Carbon Pollution Reduction Scheme (CPRS) at a time when the Australian Greens and hundreds of local climate action groups did not do so, and most other ENGOs also opposed the CPRS legislation (Roberts 2009). The Mittagong Forum, the peak organisation that allowed the environment sector some internal policy discussion and important capacity-building, went into abeyance partly as a result of this split. The state conservation councils responded by setting up their own national organisation. At the same time, hundreds of new local climate change groups are struggling to determine whether, or how, to co-operate with one another (Manning 2010; Sanders 2010), and TWS, the other main national ENGO, suffered debilitating internal ructions during 2009-2010 that limited its ability to play a strong national leadership role.

It appears that advocacy ENGOs having a community base are struggling to become the dynamic voice required on climate change, and instead the voices of climate sceptics have come to dominate the Australian debate and even 'infect' public policy. Yet, back in 1990, the Australian government was already sufficiently persuaded by environmental arguments that Cabinet agreed there was a need for a twenty per cent reduction in carbon emissions by 2005 on 1988 levels (Staples 2009). So, in twenty years, we have gone backward in the climate change debate. How has this happened and what can be done to revitalise the ENGO sector?

A strong environmental movement able to articulate its arguments and persuade public opinion, set agendas and provide vision is critical for policy success. Corporate interests and political parties can be moved by public opinion: it is the essential tool, where the sector's power lies, and it is central to a model of democracy that can be sustained for ENGOs through changes of government. I see policy failure on the part of ENGOs, and in fact of all NGOs, not in terms of specific policy issues but in terms of the relationship or model of democracy they promote and try to practise between themselves and government. This model of democracy needs to avoid co-option and be sustainable through changes of government. It is the seduction of achieving immediate gains that is probably the most powerful force leading to this failure. I will argue that understanding their role in a pluralist model of democracy is central to positioning all NGOs in a strong democratic advocacy role that will be sustainable through changes of government.

Pluralism, corporatism and public choice theory

In such a short discussion, the use of political theory can only provide a brief outline sketch. However, I will use pluralism, corporatism and public choice theory to clarify different models of the relationship between NGOs and government that have been dominant in Australia over the past twenty-five years. I also come back to classical pluralism as an ideal model that is still useful for NGOs to adopt in publicly promoting the role they play and in developing their strategy in a modern democracy. Despite much debate about the role of NGOs over the past twenty years, and attempts to refine pluralism or replace it, I believe there is still value in using the concept as an ideal for which NGOs should aim and use in public descriptions of their democratic role.

The description of pluralism most relevant to NGOs was articulated by Robert Dahl in *Who Governs?* (1961), a groundbreaking study of power and influence in New Haven, Connecticut that dissected both the formal and informal patterns of influence amongst different groups in that city using field research and historical analysis. His findings emphasised the diverse nature of power and that many interest groups compete for power in the political sphere, with the government acting as a mediator between those groups. He also articulated a dynamic model in which the relations between interests and government were continually changing and continually shifting the consensus, and in which even the political system was changing (Dahl 1961, p. 325). Dahl's model of pluralism had a significant influence on US political science, and this modern pluralism placed central importance on the role of interest groups.

Modern corporatism, which I discuss in relation to the Hawke Government, earlier had a very negative reputation because of fascist connotations, but in the 1970s it emerged as an explanation for the role of the state in modern capitalist economies (Capling & Galligan 1992, p. 42), with some theorists using it to explain the reciprocal relationships that had developed between the state and major organised interests in western counties in the post-war period. This modern version has been associated with European countries that have social democratic governments or traditions where peak associations of business and unions have worked with governments on industry policy and social welfare (Grant 1985, p. 7; Katzenstein 1985).

The origins of public choice theory, a version of which I associate with Howard government policies towards NGOs, are usually sourced to the work of US economist James Buchanan and his colleague Gordon Tullock in the 1950s. The theory has been described as:

> *... the economic study of non-market decision-making, or simply the application of economics to political science. The subject matter of public choice is the same as that of political science: the theory of the state, voting rules, voter behaviour, party politics, the bureaucracy, and so on. The methodology of public choice is that of economics, however. The basic behavioural postulate of public choice, as for economics, is that man is an egoistic, rational, utility maximizer (Mueller 1989, p. 1-2).*

The version of the theory most relevant to NGOs was that expounded by Mancur Olsen which specifically explored the role of interest groups (Olsen 1965).

This discussion will focus on ACF and the TWS as the two largest Australian national organisations covering a range of issues, but it is relevant for all ENGOs including the state-based conservation councils as well as the wider NGO sector. Other important ENGOs in Australia are Greenpeace, Friends of the Earth (FOE), and the World Wildlife Fund (WWF), but these are all international organisations that have representation in Australia. In fact, the Australian headquarters of Greenpeace and WWF manage their Australia-Pacific operations and have a focus wider than Australia, while FOE is a network of grassroots organisations but its consistent work in Australia has been primarily Melbourne-based.

Pluralism

Up until 1983, in Australia the dominant paradigm amongst scholars of business-government relations and industrial relations was 'a version of pluralist interest group theory' (Capling & Galligan 1992, p. 45), and this was the way most of us envisaged the role of NGOs at that time. Many voices - NGOs, industry and academics - all contributed to the public debate. All were legitimate voices, and the result was intended to provide better policy outcomes, help participants refine their proposals, and allow decision-makers to better develop compromises that reflected the spread of community views. Admittedly, this is an idealised vision, and in reality a business or upper-class bias often triumphed in what became known as neopluralism (Connolly 1969; Lindblom 1977; Lukes 1974; Schattschneider 1960). However, the concept of pluralism and the legitimacy of the NGO contribution were not questioned.

Mechanisms such as industry regulatory frameworks, ombudsmen, government auditors, judicial review systems and new ways to give voice to citizens, such as class actions, were developed with the aim of curbing the power of corporates and keeping pluralist voices active. But under Hawke the pluralist model of democracy was challenged by corporatism, a different model having drawbacks for NGOs, while the Howard Government challenged pluralism to its core and tried to replace it with a neo-liberal model which was a version of public choice theory and under which the NGO sector had no democratic legitimacy. The return of federal Labor governments in 2007 and 2010 signalled an end to such active repression, but ENGOs appear confused about their role. They need to be both promoting pluralism to the Australian polity and conducting ENGO advocacy guided by a pluralist paradigm.

A synthesis of the many versions of pluralism sees multiple group voices debating issues in a public sphere in order to develop some sort of consensus, and is consistent with Habermas' early articulations of the public sphere in which issues were debated in order to refine concepts and inform the political process (Habermas 1962, 1981). Definitions often include justification for NGOs, and arguments used include that they provide individuals with a stake in society, that they reflect various points of view, that they keep a balance of power amongst groups, and that they allow new voices to emerge (Connolly 1969, p. 10). Most definitions propose a public sphere in which vigorous debate refines and elaborates on proposals in a 'balancing process', but although power is shared, the final decision on implementation is made by elected representatives.

Pluralism also underpins the frequently-used Venn diagram of the three sectors of society showing market, community and state. It allows for the imprecise boundaries between them, and also explains why the NGO sector is sometimes known as the third sector.

Figure 12.1 The three sectors of state, market and community have imprecise boundaries

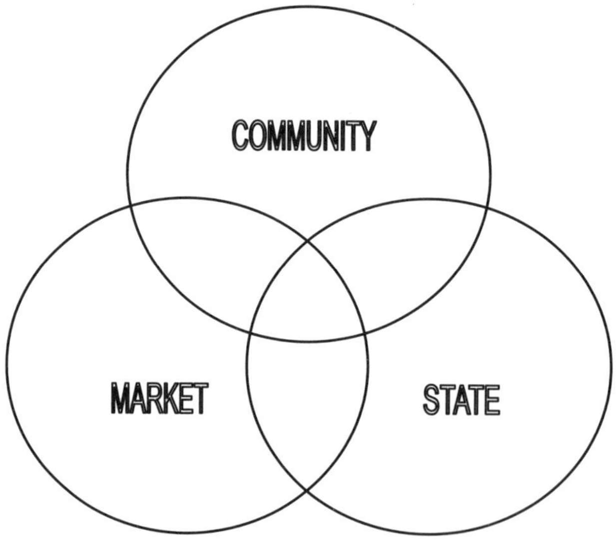

Source: Hudson 1995, p. 27

Despite its simplicity, the viewing of society in the three sectors of the Venn diagram is relevant when considering the input of each sector to the resolution of issues requiring a long-term view of policy and focused on the community interest. The state sector, or governments, is dominated by electoral cycles, so that short-term planning takes precedence. The market or corporate interests are, by their very nature, responsible to their shareholders for their bottom line of economic performance. This leaves the community or NGO sector as the voice of society that can best articulate the community welfare, look to long-term issues and provide vision. Unfortunately, this dynamic, positive view of NGOs has not always prevailed, and neither have NGOs always lived up to their possibilities. Figure 12.1, however, reinforces the idea that the NGO or community sector is different from the other sectors of market and state, and has a unique contribution to make.

Hawke government and corporatism

During the Hawke era, pluralism continued as a strong theoretical framework for describing Australian society, and it was certainly the way ENGOs approached government. But corporatism also emerged as a significant theory promoted by government. The Hawke Government's accord with the trade union movement best illustrates what Australians understand of corporatism in practice, but many Australian NGOs – such as those from social services and international development – have also created peak organisations that are corporatist in nature.

During the 1970s, theorists described a form of corporatism to explain the reciprocal relationships that had developed between governments and organised interests such as unions and industry groups in western countries in the post-war period, describing it as:

> *... a system of interest representation in which the constituent units are organised into a limited number of singular, compulsory, non-competitive, hierarchically-ordered and functionally differentiated categories, recognised or licensed (if not created) by the state and granted a deliberate representational monopoly within their respective categories in exchange for observing certain controls on their selection of leaders and articulation of demands and supporters (Schmitter 1974, p. 93–94).*

Interest intermediation is one of the key characteristics distinguishing pluralism from corporatism, and it is an important factor limiting the democratic advocacy role of organised labour and NGOs under corporatism. It does this by giving power or recognition to peak groups that may hold different policy positions from those of individual groups. This power allows a peak group to enforce compliance with corporatist agreements that will stifle the individual groups holding minority positions within the peak. It fuses or limits the multiplicity of policy positions that may be found in its member groups, and this is most likely to be fusion resulting in a weaker policy outcome than some members would hold.

The environment sector has never had a peak group organised in this way. I was repeatedly urged by Graham Richardson, the Hawke Government Minister, to encourage ACF to form a peak environment organisation for Australia and, after I left, my colleagues Judy Lambert of TWS and Philip Toyne of ACF experienced the same pressure (Lambert 2008; Toyne 2008). The model promoted to us by Richardson was that the environment organisations needed a group 'like the ACTU' (Australian Council of Trade Unions) (Lambert 2008). None of us was interested in promoting this model. We saw a peak group as requiring organisations holding strong policy positions to accept the lowest common denominator put forward by all; as requiring much energy and resources to organise, which we judged could be better spent on advocacy; and as weakening the pluralism of many different voices promoting a variety of policy positions. Despite a number of other NGO sectors choosing to form themselves into corporatist peaks, we rejected such a model. More recently, the environment sector has come up with a creative new model that overcomes the disadvantages of a corporatist peak and I will discuss this later.

Corporatism played a part in the relationship between ACF/TWS and the Hawke Government. Hawke and key ministers saw the endorsement of Labor by ACF and TWS at the 1983, 1987 and 1990 elections as being a corporatist arrangement. This can be seen in the many complaints of ministers that, despite the Government's agreement to various policies, the environment organisations 'would never be satisfied' (Balderstone 2008; Emerson 2008), whereas the environment organisations held a pluralist view and continued to push for their claims to the maximum. For example, at the 1987 election ACF and TWS insisted on the implementation of the three major issues they had been pursuing over the previous terms of government: an end to logging of Queensland rainforests and World Heritage status for the wet tropics; action to stop logging of Tasmania's forests; and completion of the Kakadu National Park by the addition of Stage III with the whole of Kakadu becoming World Heritage. In addition to this ambitious agenda, they also insisted on the government opposing a Japanese company being allowed to mine at Shelbourne Bay on Cape York Peninsula, despite Treasury advising that to do so threatened Australia's international trade reputation (Emerson 2008). Much of the tension between the Hawke Government and ENGOs at the time was because of an unarticulated difference in how the two sides saw the relationship between them.

In exchange for endorsing Labor, ACF and TWS were successful in having the Government agree to all of their proposals. By 1990, other parts of the environment sector were complaining that ACF and TWS had moved too far away from a pluralist relationship by repeatedly supporting Labor. Friends of the Earth (FOE) feared that ACF and TWS were putting the environment movement in a position whereby they could be taken for granted and become part of a corporatist relationship. In their criticism of ACF and TWS, FOE said the strength of the sector was in it being 'a motley collection of groups and individuals, and much to the annoyance of some politicians and sections of the media, can not [sic] be labeled and pinned down' (O'Loughlin 1990, p. 19). After endorsing Labor for three elections, the relationship was in great danger of morphing into one of corporatism and co-option, as inevitably the support of the organisations would be taken for granted by Labor, as Hal Wootten warned when he was President of ACF in 1987 (Wootten 1987).

Toyne and Balderstone describe the Hawke Government as achieving more to protect the environment than any government before or since, and they list eleven very significant environmental achievements (Toyne & Balderstone 2003). The magnitude of the progress achieved through the relationship with environment organisations is not disputed, but Alec Marr, former Director of the Wilderness

Society, identified the main negative legacy of the period as being that many in the environment movement believe that Labor has a natural affinity with ENGOs (Marr 2008). This is despite the fact that Labor turned its back on the environment organisations under Keating, removing forestry as a Federal government responsibility, ignoring the Cabinet decision on climate change, dropping the ecologically sustainable development process, and failing to ensure that Keating, as Prime Minister, attended the 1992 UN Rio Earth Summit on the environment. Interviews with NGO leaders support Marr's contention. Time and time again, they quote significant occasions when Keating was not just *uninterested* in the environment, but actively antagonistic towards the sector (Lambert 2008; Parlane 2010; Marr 2008). The Rudd Government's failure on climate change legislation would seem to add weight to this thesis (Crowley, this volume). Thus, the Hawke legacy may have created a bias in the ENGO sector towards Labor that is not founded on historical facts. This reinforces the importance of a strictly non-partisan stance towards political parties – a stance that is inherent in pluralism.

Howard Government and neo-liberal public choice theory

With the accession of the Howard Government, we saw the articulation of a new form of the relationship based on neo-liberal public choice theory that resulted in serious attacks on the legitimacy of the wider NGO sector and tried to stop their public advocacy. I have written in detail elsewhere on how closely the Howard Government's actions towards the whole NGO sector were linked to the logic of public choice theory (Staples 2006, 2008).

Public choice theory applies marketplace principles of human behaviour to the public political sphere. It assumes that people's political behaviour is influenced mainly by their rational self-interest, and that the motivation of people in the development of public policy is no different from their motivation in a commercial market environment. Public choice theorist, Mancur Olsen, argued that interest groups or NGOs arise in order to gain advantages at the expense of the many (Olsen 1965). His analysis did not incorporate the complexities of NGO motivation, including altruism, nor that NGOs are able to take a long-term perspective – even past the lifespan of individuals or the life of the interest group. The theory is linked to a narrow, classical view of representative democracy that holds that only *elected* representatives have responsibility for public policy (Mueller 1989).

The shift in public discourse introduced by the Howard Government indicated a new hegemony of public choice with references to the mainstream, to ordinary Australians, and to the Australian way, coupled with criticisms of special interests, of single-issue groups and of elites, and with an emphasis on accountability. An emphasis on the lack of accountability of advocacy organisations was a consistent theme of the Howard Government and neo-liberal think tanks such as the Institute of Public Affairs (IPA). Many interpreted this as a criticism of NGOs' internal governance, but it was actually a reflection of the view that NGOs are not legitimate players in public debate because they are not elected representatives (Staples 2008). Throughout the Howard years, the logic of the theory was implemented by a series of actions aimed at stopping the advocacy of NGOs, and after the Government gained control of the Senate in 2004 the attacks on NGO advocacy became even more pronounced. These attacks have been documented by a number of authors (see Hamilton & Maddison 2007; Maddison & Denniss 2005; Manning 2004).

The legacy of the Hawke and Howard years

The extremes of public choice repression were lifted following the change to the Rudd Labor Government in 2007. Confidentiality clauses preventing NGOs who received government funds speaking to the media were removed. Julia Gillard, in her role as Minister for Social Inclusion, said very early that the government would 'get the gag off and listen to those who know what's going on' (The Australian 2007, p.1). After being shut out for a decade, ENGOs such as ACF found that they now had access to ministers (Tupper 2008).

The end of the Howard era had resonance with the ending of the Thatcher period in the UK where Prime Minister Thatcher implemented a repressive regime against NGOs using policies consistent with public choice. John Dryzek and various overseas colleagues (Dryzek *et al.* 2003) examined the relationship of the environment movement and government in the United States (US), the United Kingdom (UK), Germany and Norway over a twenty-year period. They reported that, in the UK, when the Thatcher Government came to an end, the environment groups were relieved that the repression was lifted, but they did not again become strong advocates for their causes. The major negative effect of this active repression was 'in just how little' environment groups were 'prepared to settle for once it was lifted' (Dryzek *et al.* 2003, p. 124). This situation that occurred in the UK should be a warning to Australian NGOs of the need to reassess their strategic position in relation to public advocacy and to ensure that, with the lifting of active repression, they are acting as vigorous public advocates in the public interest. For example, ACF has been criticised by others in the sector for taking a compromised stand on climate change in May 2009 by supporting the Government's amended Carbon Pollution Reduction Scheme (CPRS) (Roberts 2009). I believe that ACF misunderstood their relationship with government in taking this position They did not need to compromise and reference to the pluralist model would have helped clarify their direction.

Although electoral endorsement of political parties is not contemplated by NGOs these days, it is not inconsistent with pluralism but does run the danger of changing toward corporatism once such a relationship becomes entrenched with the same party. Today, instead of electoral endorsement, various sophisticated methods of assessing party policies are used to inform voters and this is totally consistent with pluralism. However, there are still ENGOs acting outside elections as if their endorsement of the policies of political parties can create gains for them. Yet any endorsement runs the danger of weakening ENGOs. I do not exclude encouragement and recognition when political parties propose good policy, nor do I exclude an ENGO being pragmatic in developing its political stance. Different positions are part of a pluralist model. But I am interested here in how ENGOs use the public sphere and that they should not over-reach themselves in thinking they can decide policy. If we refer again to pluralism, ENGOs do not have the power of elected representatives and it is the political parties' role to compromise on the many points of view put forward under pluralism. If they are confident in their policy, then ENGOs can be confident that it is not their role to weaken their position as they have no power to implement policies. On the issue of climate change, I see that their role is to represent the science as honestly as they can. It is the role of political parties to decide where they will compromise on implementing policies that address the scientific imperatives we face.

Another legacy of the past that I believe has not been fully integrated into ENGOs' strategic behaviour relates to the entry of the Australian Greens into national politics. This first occurred at the 1993 election and at a time that electoral endorsement was not envisaged by the sector because of Keating's position on environmental issues. Chris Puplick, former Shadow Coalition Environment Minister, told me he saw the entry of the Greens as leading to a reduction in the influence the environment organisations had with the other political parties, because the Greens provided a natural home for the environment vote (Puplick 2008). For Labor, obtaining the second preferences of the Greens became far more valuable than any influence or endorsement that the ENGOs could offer. Negotiating preferences with the Greens became a most significant imperative for political survival, than developing corporatist electoral and policy arrangements with ACF and TWS. From the point of view of the environment organisations, it was impossible to recommend a primary vote for one of the major political parties when the Australian Greens' policies were so close to their own. At the same time, to recommend a second preference vote for the Greens as they did with the Democrats in the 1990 election, would cut across any preference deals negotiated by the Greens. Puplick emphasised to me that he believed the rise of the Greens was 'an absolutely critical factor' in the changed relationship that ENGOs have with the other parties (Puplick 2008). I am not sure that the ENGOs have yet fully appreciated this fact.

This can be seen in the environment sector's positioning in relation to the Greens, and that ENGOs do not need to weaken their positions but that political parties do. The pluralist model suggests that, where there is a sympathetic political party such as the Greens, if they agree with the direction of that party's policy, then ENGOs should not propose policy positions that weaker the negotiating ability of that party. The Greens are becoming increasingly powerful. As I write, they have a representative in the House of Representative helping to keep the Labor Government in power, and from July 2011 they held the balance of power in the Senate. The Greens have to compromise their positions in voting on issues, as was seen in countless issues including the climate change legislation. The Greens are elected representatives whose role in a pluralist model is to compromise in finding a way through the many competing policy proposals that are politically possible. The ENGO sector's role under pluralism is not to weaken their policy in this way, because the sector is not the decision-maker that has to vote on the issue.

In this context, the sector should be holding a much stronger position on climate change than the Greens in order to allow the Greens to do the compromising. Unfortunately, this has not been the case for some key ENGOs, with dire policy consequences.

Where NGO power lies

There has been some research that suggests NGOs' strongest advocacy position is to work on public opinion rather than from the position of 'insiders'. As a former ACF Canberra lobbyist, I do not suggest that engaging with government is not appropriate, rather I am interested in NGO positioning that will optimise democratic advocacy and will be sustainable through changes of government. The international study referred to earlier by Dryzek *et al.* (2003) looked at the relationship between the environment movement and government in the US, the UK, Germany and Norway over a twenty-year period. The researchers found that the environment movement in Germany was the most successful in changing the core values of the state towards environmentalism, because the state practised passive exclusion towards them. Government policy set this position, not NGO strategy, but it showed that the value of staying outside government processes and using public opinion was found to be the strongest model for influence despite the rise and the sustained successes of green politics in Germany. Maddison and Edgar (2008) come to a similar conclusion with different theory using the examples of the Franklin Dam campaign, and of ENGOs in Queensland who chose to stay outside government processes on the issue of land clearing, despite pressure from government to engage in a consultation process. They conclude that there is great value in mobilising public opinion to influence governments rather than engaging through consultation and committees.

Choosing when to engage and when to stay outside is not a simple issue. With the rise of environmentalism, there are many requests by companies and governments to join committees, provide advice, play consultative roles, provide submissions, and so on. This can be very flattering to organisations that have previously been outsiders. Although tempting, accepting such offers can dilute the energy of ENGOs and deflect them from their role of setting agendas in the public sphere. Balancing both engagement and staying outside is the biggest challenge, but the pluralist model can assist with this choice. The conclusions of Drysek *et al.* (2003) and Maddision and Edgar (2009) point to the need to remember that the power of ENGOs lies in their ability to persuade the public to their cause, and to argue persuasively in the public sphere. Their strength is in shaping public opinion, setting agendas and providing vision.

This strength is also the path that leads to long-term sustainability in ENGOs' relationships with all sides of the political spectrum. Governments change, and it is important that ENGOs practise pluralism and advocacy that is totally independent of any political bias. ENGOs' embrace of pluralism should not just be in its practice, but also in its promulgation. ENGOs must educate and persuade the public, hence decision-makers, of the important democratic role they have to play in an effective, inclusive democracy. The hegemony of public choice theory under Howard has not totally disappeared. Indications are that it would have returned were a Coalition government to have been elected in 2010. Apart from John Howard, two of the main NGO antagonists - namely Senators Abetz and Mason - are still active in Coalition ranks in Parliament, with their views of NGOs well documented (Abetz 2004, 2006; Mason 2004a, 2004b). Senator Abetz is Shadow Minister for Employment and Workplace Relations as well as Coalition Leader in the Senate, and Senator Mason is Shadow Minister for Universities and Research. Persuading the public of the value of a pluralist model of democracy and the importance of the NGO sector is significant work that must not be neglected. There is promise in a parliament where independents and Greens have policies consistent with this aim, but ENGOs cannot leave the selling of pluralism and their important role to others. This should be part of ENGOs' core business. The government will eventually change, and a defence is needed to be built into public opinion. Unless ENGOs understand this and act upon it, they will be guilty of policy failure at the very least.

A co-operative alliance

A positive policy response by the environment sector during the Howard years was to develop the Mittagong Forum. This organisation represents an innovative model that is outstanding in providing communication, co-operation and capacity-building without having the disadvantages of corporatism found in a traditional peak body. In fact, it is a model that would be useful for all NGO sectors to emulate. It is unfortunate that the Forum is currently in abeyance. Formed in 1999, it was created by bringing together the major national groups, the state conservation councils and specialist issue groups. Unlike traditional corporatist peaks, such as the Australian Council of Social Service or the Australian Council for International Development, it did not develop joint policy or provide a single voice to government from the member organisations; neither did it maintain staff other than one part-time co-ordinator, and it did not undertake research unless by collaboration amongst members. The model of not undertaking direct public advocacy was the key that set it apart from corporatist peak groups. In every other way, it fulfilled functions familiar to a peak organisation. Soon after its formation, the Forum undertook a major capacity-building exercise over a two-year period, with its member groups focusing on governance skills aimed at improving member capacity. Although it did not develop policy to present to government on behalf of the sector, regular meetings provided an opportunity for the groups to share information on government policy in relation to the environment and ENGO governance. These meetings included comparing and discussing their responses, deciding on sharing tasks in relation to mutual concerns, and hearing from experts in various fields relevant to government policy (Alexander 2010). Its particular value was that it was intended to provide a place for reflection on strategy, for nurturing leadership, and for capacity-building across the environment sector (Alexander 2010; Fisher 2010).

The immediate factor causing the Forum to suspend operations was ACF breaking with the rest of the sector when it supported the Rudd Government's climate change legislation. At the same time, the Forum's financial support had narrowed from its original broad base, so that ACF's assistance both in kind and in finding donors had become essential to its operation (Alexander 2010). But there were more basic factors that Russell Fisher, President of Environment Victoria, and Karen Alexander, long-time environmental leader, have identified. Firstly, there was continuing tension between the attitude of political campaigners focused on political outcomes, and the need for environment personnel 'to stop and take time out for reflection, for development and for nurturing of leaders'. Second was the fast turnover in personnel in the environment sector, while thirdly, there was continuing tension between national, regional and state-based groups that have different levels of resources upon which to draw and different priorities (Alexander 2010, Fisher 2010).

The Mittagong Forum obviously had significant obstacles that led to its demise, but Fisher claims that they continue to be obstacles to co-operation within the sector that are not being addressed. Since the Forum has been in abeyance, the state conservation councils have formed a peak group, and Fisher identifies similar difficulties and tensions in *its* operation. In particular, he points to 'the emphasis on campaigning, not capacity-building for the sector'. He claims that it is 'politics that energises it, not nurturing personnel or the sector', and that for campaigners, the new conservation council peak group 'loses relevance for campaigners if it does not do policy' (Fisher 2010). Unlike the Mittagong Forum, the conservation council peak is not a sector-wide peak, as it does not include ACF, TWS or the large international organisations such as Greenpeace and WWF (Alexander 2010). So, the obstacles continue to having a co-operative alliance that both nurtures the environment sector while allowing for policy discussion and enrichment. Despite these tensions, the model is one that has significant advantages over a traditional peak because it reinforces a strategic pluralism without sacrificing the individual independence of members into a corporatist organisation. The environment sector would do well to reconsider its formation while addressing the problems that led to its demise.

Conclusions

A strong environmental movement able to articulate its arguments and persuade public opinion, set agendas, and provide vision is critical for the adoption of environment policy by governments. But policy capacity on the part of ENGOs should not be measured simply in terms of specific short-term successes and/or failures. Instead, it should be measured against whether they have been able to influence policy within a framework that will be democratically sustainable so that the sector will have continuing influence. Specifically, ENGO policy failure lies in not consistently promoting and practising advocacy in a pluralist model of democracy that avoids co-option and is sustainable through changes of government. The history of the past few decades suggests that the seduction of achieving immediate gains and an inability to recognise different models of democracy have contributed to a failure on the part of many Australian ENGOs.

A fresh analysis of their position in relation to the government is required in order to move out of the shadow of past history – that is, of endorsing Labor under Hawke, and the experience of repression of advocacy during the Howard years. Understanding their role in a pluralist model of democracy is central to positioning ENGOs in a strong democratic advocacy role that will be sustainable through changes of government. This model clarifies the extent to which ENGOs need to weaken their positions during negotiations with government, and in relation to minor sympathetic parties such as the Greens. Climate change is a case in point where ENGOs would do best to promote the recommendations of scientists and let the political parties make any compromises on the final policy adopted by Australia. This is particularly relevant as the Australian Greens are increasing their political power and able to play a key role in any government negotiations.

ENGOs should also adopt and explicitly promote the model of pluralism and their role in it at all times. This should not be as a specific separate campaign that waxes, wanes and then disappears. Instead, the model and the role of NGOs should be referred to and endorsed as opportunities present themselves in every environmental campaign. The reinvigoration of the Mittagong Forum would be an immediate step towards strengthening the sector. The difficulties that saw its demise are worth overcoming. The benefits of co-operation without coercion, and increased skill-building within the sector would be considerable. But the overall aim for ENGOs should be to ensure that public understanding and support for a pluralist model of democracy will extend through changes of government and will provide a defence should a government come to power that attempts to promote policies consistent with public choice or undermine the democratic advocacy role of ENGOs.

Addressing climate change and environmental sustainability are core issues that the world now faces. In Australia, we need vigorous ENGOs setting agendas and remaining strategically positioned regardless of what government is in power. It remains to be seen whether our environment movement can put past failures behind it and rise to the challenge.

Chapter 13

Conclusions

KATE CROWLEY AND KJ WALKER

Introduction

Over a decade ago, Yencken argued that not only is Australia's ecological decline persisting, but that it is intensifying despite all our environmental policy efforts thus far (Yencken 2000, p. 4). This is surely policy failure of the highest order, and evidence that ecological concern is not yet integrated in any meaningful way into decision-making. It is certainly possible to point to positive environmental policy progress over the past several decades, but primarily in the sense of institutional innovation, and not in the sense of innovation making any great environmental difference. There is frustration, too, with flawed policy implementation, with lack of resourcing, with inadequate monitoring, with poor evaluation, and with neglected or abandoned initiatives.

Above this level of policy detail, we can begin to make sense of the systemic nature of the problem as we appreciate the implications of clashing values, environmental and developmental, and the persistent over-emphasis on natural resource exploitation. The political priorities of economic growth, industry certainty and the intergovernmental relations that facilitate these priorities very often work against strong environmental policy action. The efforts that were made in the 1990s to overcome this tension, to pursue ecologically sustainable development, and to integrate environmental concern have almost certainly failed, although the sentiment lives on in policy preambles and rhetoric.

There have been few policy reviews in this field. The most extensive is Dovers and Wild River (2003), in which Dovers reflects upon the efficacy of Australian environmental policy processes and institutional arrangements over the past few decades, offering comprehensive insights into the problems of policy failure. In terms of environmental policy persistence, he notes that many policies and institutions are not supported long enough, nor are they sufficiently robust, to make any difference. Lacking too is any purposefulness or agreed principles, goals or standards for environmental policy processes and institutional arrangements. Nor is it clear that there is sufficient information richness and sensitivity upon which to base policy and management practice, with basic environmental monitoring unclear at best, and long neglected or lacking altogether at worst.

Inclusiveness has been urged as a criterion for reflection about environmental policy processes and institutional arrangements. For a variety of reasons, it is found to lack clarity. Firstly, Australia has seen a remarkable increase in community-based management and monitoring programs, but conversely it has also seen a trend towards the devaluing of participation. This has included, for example, a diminution of public standing in environmental law in Australia. Participative processes were found in the Dovers-Wild River study to suffer from poor design and a lack of genuine commitment by governments, but there was also little evidence found that improved or enhanced public participation would markedly affect or improve environmental policy outcomes (Dovers & Wild River 2003, p. 524). There is no compelling record of environmental policy innovation or institutional flexibility. In the face of new and emerging information, adaptive capacity, or the reflecting upon lessons learnt from earlier experience, is strikingly deficient. The test of adequacy is simply not met.

In reviewing some of Australia's most critically challenging environmental issues and assessing the capacity of contemporary policies to solve them, measured by the key indicator of ecological improvement, this book reveals, case by case, the nature of environmental policy failure in this country. It is characteristic, for example, of both Labor and conservative governments across many generations. But this critical failure is so much more than a mere failure of political will, although this has been central, for example, to the inability of Australian federal governments to act decisively on climate change. Rather, it is deeply rooted on many fronts, including: Eurocentric development practices, short-term thinking, self-interest and corporate power, and the neglect of ecological sensitivities and sustainability, which combine to thwart effective policy development.

Now time is running out in a country where the natural environment is more vulnerable to impacts than in many other developed countries, and where there is a long-standing record of policy failure steeped in politics, neglect and incompetence. The value of ecological capital is yet to be appreciated in the 'quarry economy', as natural resources belonging to the public domain are undervalued and oversold, and where the rate of ecosystem destruction is unparalleled in the world. Each of our studies here considers environmental policy legacy in terms of past efforts and effectiveness, current critical issues and challenges, and implications for future policy. As a whole, they find that not only is the environment as a natural asset being irreparably degraded in Australia, but that the intrinsic value of its natural heritage is being eroded and the integrity of ecosystems is under serious threat.

Australian stories

Walker takes up the story in *Chapter Two* by considering the very many obstacles to long-term, transformative environmental policy regimes which must be adopted if the ecological damage wrought thus far is to be reversed. He finds that the dominant policies are exploitative, and such palliative measures as do exist fail to take account of ecological carrying capacity and limits to natural resources; this in a country with a persistently 'developmentalist' policy context where the state is deeply implicated. In *Chapter Three*, Crowley finds palliative policy alive and well in the climate change sphere. She describes the failure of the Rudd Labor government to implement its own Carbon Pollution Reduction Scheme (CPRS), a scheme that was to differentiate it from the previous government's failure to act on climate change. In fact, the criticism of this scheme was not only that it would have imposed a new cost across the board, but also that it would have perpetuated, not prevented, carbon pollution and so offered only symbolic climate action at best.

In *Chapter Four,* Mark Diesendorf drills down more specifically into climate policy and identifies a lack of transformative energy policies and thus a failure of successive Australian governments, conservative and liberal, to respond to the twin threats of climate change and energy security. Industry lobbying has not only thwarted the CPRS, but has also influenced a remarkable neglect of renewable energy and energy efficiency, the only genuinely low-carbon technologies and measures capable of achieving large energy emission reductions by 2020. Moving on to natural resources management, in *Chapter Five* Geoff Cockfield describes the unsuccessful balancing of the contrasting environmental and economic management approaches and the inevitable parallel decline in biodiversity. Here, his story parallels the atmospheric decline and climate impacts that are being linked globally, as well as in Australia, to climate change inaction. Cockfield finds that an unstable accommodation of environmental and economic ideologies by governments has produced weak policy and governance, and timidity about acting at the critical inter-regional scale.

There are related problems with the management of the Murray-Darling Basin, examined in *Chapter Six* by Daniel Connell. In this case, policy weakness is in part the result of lobbying by developers who have succeeded in undermining the effectiveness of institutional, intergovernmental and policy efforts. The politicisation of climate policy, energy policy, natural resources and water management to the detriment of effective environmental policy with improved ecological outcomes thus emerges as a strong theme in each chapter so far. Furthermore, policy failure in these arenas is longstanding, and therefore all the more acute. Climate and energy policy, for example, have been on the agenda since the 1988 Toronto World Conference on the Changing Atmosphere, and the management of the Murray River has of course been debated and attempted for over 100 years.

In *Chapter Seven,* Wescott finds that twenty years after the landmark Resource Assessment Commission's Coastal Zone Inquiry, there is still no clarity about the federal government's role in coastal planning and management. Here, again, policy failure is longstanding. Nor is there clarity about oceans policy, with the innovative Australian Oceans Policy dormant, without legislative support, and with its ministerial, advisory and institutional governance arrangements all dismantled. Similarly, Michael Buxton finds in *Chapter Eight* that many decades have passed since political leaders recognised serious issues with the nation's system of water allocation, but still there is no clear solution in sight. Recent efforts to establish market-based water allocation mechanisms have been fraught with failure, and indeed are repeating the longstanding regulatory failures of misallocation, over allocation and accounting inadequacies. The complexity of politics between levels of government helps to account for the palliative policy response in coastal and water management.

Indeed, the mismatch between Australia's complex and interrelated environmental problems and its late nineteenth century governing institutions is a recurring critical theme in this book.

In *Chapter Nine*, Michael Howes and Aysin Dedekorkut-Howes find that the emergent challenge of planning for climate change adaptation is all the more difficult given limited public resources, competing demands and political power struggles. The frustration of climate change mitigation policy by 'short-term politics, political rivalries and a persistent campaign of opposition by sceptics supported by a powerful fossil fuel sector' found by these authors does not bode well for adaptation policy. Climate change adaptation policy will eventually be forced upon governments as the consequences of spiralling demands for mitigation hit home, but the message from this chapter is that Australia's political institutions are poorly equipped to respond.

This theme of political and institutional preparedness for responding to environmental disaster carries over into *Chapter Ten* as Dave Mercer considers the politics of catastrophic bushfire and the public interest. He finds evidence that the risk of catastrophic fire danger will soar by 2050, based on the current trajectory of greenhouse gas emissions, but that preparedness is hampered on a number of fronts. There is the lack of policy learning after previous catastrophic events, and failure to aggressively reduce the vulnerability of future targets, coupled with, indeed aggravated by, short-term political thinking. The persistence of policy failure is again appalling as Mercer observes: 'what is alarming is that so little appears to have changed in terms of planning, coordination and response strategies' in the 70 years after the 1939 Bushfire Royal Commission. Long-term, systemic policy neglect, already identified on so many fronts, features in *Chapter Eleven* as Mark O'Connor argues that Australia's missing population policy is perhaps its most serious neglected environmental issue.

All too often analysts claim that the remedy for environmental policy dilemmas, neglect and disasters is enhanced democratisation, greener politics and public participation. This is often the cure-all, end-point of their analysis. There are indeed many senses in which enhanced democratisation could contribute to improved environmental policy: in terms of climate, water and natural resources policy, for example, by balancing the influence of powerful interest groups. We have seen private developers and corporate influence dominate in these arenas, and some balancing institutionalisation of intrinsic and community-based environmental interests may bear consideration. However, in *Chapter Twelve* Joan Staples offers a revealing cautionary consideration of the adequacy of Australian environmental non-governmental organisations (ENGOs) themselves in terms of the insider roles they have assumed over time. In her terms, it is simplistic and ill conceived for ENGOs to allow themselves to be institutionalised when what is needed are independent voices that are stridently critical and independent of government.

Studies in policy failure

How then to critique and explain these studies in policy failure? The discussion in *Chapter One* suggested that there are four very broad lenses of analysis that could be employed when discussing policy success or failure: context, process, programs and politics. Context was the criterion that we added to account for ecology, which was described as the primary indicator when considering environmental policy success. The sorts of issues we could discuss under the context heading include the complex character of ecology, the wicked challenge of environmental policy, the nature of the environmental problem to hand, and whether the problem was solved. We would expect to see environmental quality improve, not decline, for there to be policy success. As Walker suggests in *Chapter Two,* government policy is adequate when it can be shown to have identified a problem in a timely fashion and to be tackling it at a rate sufficient to stem it, reverse it, or at least manage it. This is policy success.

In these terms we could judge a drop in riverine pollution, an increase in protected old-growth forests, a return to estuarine health, a decline in threatened species and so forth as policy successes where there are programs to achieve these ends. This study has not considered localised programs, but has looked more broadly at Australian environmental issues and interrogated national environmental policy failures. It has chosen to look at issues where greenhouse gas emissions are soaring, where renewable energy is neglected, where biodiversity, the Murray-Darling Basin, coastal and ocean health, and water quality are all declining. Given such historical policy failures, the study then moves on to question whether Australian society is prepared to deal with the consequences, for example, of climate change impact and catastrophic bushfires. Not only are there systemic environmental policy failures, it finds, but also chronic institutional and political failings in dealing with the very real social and economic consequences that can follow such failures. Recent natural disasters, and the immense social and economic pain that these have caused, have confirmed these as costs that can only be avoided for a limited time by failing to factor in environmental aspects to political and economic decision-making.

By moving on from policy context to considering policy process, programs and politics, this study seeks to explain how such failures occur and how they may be overcome.

We saw in *Chapter One* that policy process is not just a study of the mechanics of policy design and implementation, but also a consideration of legitimacy, relevance, capacity and political sustainability. For example, the CPRS may have been intended to be well designed and well implemented, but it was rejected as illegitimate by critics, not least for failing to target emissions reductions. Renewable energy policy also fails the policy process test as a neglected option with no legitimate, relevant or capable dimensions. Natural resource management (NRM) and national oceans policies greatly lack capacity, again partly through neglect but also through failed implementation that could be described as policy abandonment. Many of the policy processes studied—NRM, coastal and water policy processes, for example—fail through lack of clarity of purpose and a lack of integration across sectors, interests and agencies, but notably also across levels of government.

Probably the most astounding failure is the failed management of the Murray-Darling Basin (MDB), given the one hundred years that it has been on the agenda, and its steady environmental decline despite all manner of policies, institutions and processes. The MDB study shows how very difficult it is to disentangle policy process from programs and politics. Numerous very rational, well-resourced and well-informed efforts at improving the basin's management have been thwarted by developers and interest group pressure, and by short-term politics. But this study highlights, in particular, the failed role of national government in intergovernmental environmental arrangements, illustrating Howes and Dedekorkut-Howes' point that Australia's governing institutions are a product of nineteenth century society and simply not equipped to deal with managing the environment.

What of Australian environmental policy programs? These, as we saw in *Chapter One*, need to be purposeful, operational, well informed and resourced, integrated, lasting, well embedded in statutory terms, and transformative not palliative. Climate change programs appear to have been palliative, hit and miss, neglectful of emissions reduction, and in some cases to have been abandoned, whilst transformative programs have been shunned altogether. NRM programs may have been well intended and well funded, but their failure has been assured because they are under-utilised and contradictory, with management plans bearing little relationship to broader objectives. It is hard to say that there has been a coastal or ocean policy, let alone a full-blooded program, whilst water policy programs seem to have narrowly focused on a failing market-based approach. Adaptation policy and programs are struggling into being, whilst bushfire preparedness seems altogether captive to short-term political decision-making.

Even in the light of the limited coverage offered here, it is not difficult to see why much of Australian environmental policy analysis simply focuses on politics. Where there is no rational policy response, where policy is undermined or abandoned, where private interests dominate and where environmental problems are neglected or displaced, politics is never too far from the scene. We saw in *Chapter One* that, for an environmental policy to be considered a political success:

it must be politically popular; it must have enduring support; and it must pursue sustainability. Ideally, political leaders would show determination, would integrate the environment into their decision-making, and would even codify the goal of sustainability into statutory frameworks. And yet we have recently seen that an unsustainable and misdirected environmental stance may achieve political popularity with affected parties and corporate interests, while at the same time actually undermining a more ecologically rational policy.

Whilst the role of ENGOs should be eternal public vigilance to counter corporate pressures that undermine political will for ecologically rational action, in particular, Staples has shown in *Chapter Twelve* that this is very clearly not always the case. Australia's peak ENGOs are failing where they succumb to co-option—essentially, Staples claims, by Labor governments with dire consequences for both democracy and environmental policy. The consequences of quiescent ENGOs and captive government can be overwhelmingly disastrous.

In *Chapter Three*, for example, eminent economist Ross Garnaut credits vested corporate interests with derailing the Rudd government's proposed emissions trading scheme (ETS). In *Chapter Four*, the fossil fuel industry is credited with the policy neglect of renewable energy. In *Chapter Five*, development pressures are seen to be undermining attempts at natural resource management. In *Chapter Six*, again the dominance of developers is credited as a major cause of the failure of Murray-Darling Basin reforms. *Chapter Seven* finds the neglect and unravelling of coastal and marine policy favouring the oil, gas and fishing industries and interests over and above scientific and conservation concerns. In *Chapter Eight* we see that failing market-based water policy crowds out ecologically sensitive solutions, to the advantage of articulate business-based interest groups. There is less corporate influence in adaptation policies and disaster response in *Chapters Nine* and *Ten*, but short-term, political interests are still dominant. And the so-called population 'boosters' criticised in *Chapter Eleven* include prominent media, business and industry-based lobby groups.

Reluctant nation?

It is critical in considering how environmental policy failure may be overcome in Australia today to revisit the 'reluctant nation' thesis. Toyne (1994, p. 7) holds that '[d]espite the generally expansive interpretation of the Commonwealth's constitutional power, there has been a reluctance to interfere with the State's Rights *status quo* generally and little evidence that a national government will act consistently and decisively to set a national approach to managing the environment in the national interest'. There is no environmental head of power in the Constitution, however the Commonwealth has established many other valid avenues for legitimate intervention in environmental affairs, as we saw in *Chapter Nine*.

Each chapter here has found that the federal government is a crucial element in environmental policy development and management efforts. A discussion of the federal dimension, therefore, completes the consideration of the political context of environmental policy success or failure.

Chapter by chapter, the study confirms that the reluctant nation thesis is alive and well today on a variety of fronts, with no consistent or decisive action by the federal government and crucially no *bone fide* federal move towards ecological sustainability.

In *Chapter Two,* Walker finds that, since 1788, Australian public policy has been persistently developmentalist, with government subsidising economic activity to promote development under a régime he calls *statist developmentalism.* From the 1980s onwards, there was a shift at the federal level towards economic rationalism to the detriment of previously promising environmental policy trends, but this had very little impact on the entrenched developmentalist paradigm. This national trend of 'economically rational' developmentalism deepened from the 1990s during the years of the Howard conservative government which was strongly neo-liberal. Unplanned economic growth and expansion in keeping with the historical 'quarry Australia' mindset were emphasised, whilst the long-term consequences and ecological impact were ignored. The consequences of this are clear to see here in each chapter.

As we saw in *Chapter Three,* when the Rudd federal government decided to act on climate change it was within this context, which explains both its blind faith in an ETS market mechanism and its watering down and abandonment of the ETS under pressure from industry. As noted in *Chapter Four,* the Rudd government failed to lead from the national level on promoting transformative energy solutions, but instead enhanced the dominant role of fossil fuels while restricting renewable energy very much to a niche role. This interventionist federal government behaviour has very actively protected vested interests, contrasting sharply with more reluctant behaviour on other environmental policy fronts. The federal government only took an interest in natural resources management from the 1980s when it introduced Landcare, which has evolved as a collaborative 'caring' model into further and grander iterations. Here, the federal government acts as facilitator through funding agreed devolved actions, producing a form of bioregional governance that Cockfield describes as weak in *Chapter Five.*

In natural resources management, the federal government could control policy more effectively using its bilateral funding agreement lever. The solution to more effective management of the Murray-Darling Basin is more complex. This case study in *Chapter Six* demonstrates the constitutional and institutional constraints upon federal government action across the basin where its authorities have only advisory and discretionary responsibilities. The 'unanimity' decision-making principle and the self-interest of the affected states mean that decision-making is highly politicised and that the federal government may need to pursue a regulatory intervention option in water management. This is at least being considered, unlike any resolution to the lack of national coastal policy and planning and the failed condition of Australia's much-lauded marine policy, as noted in *Chapter Seven.*

Indeed, without establishing a viable market-based funding, political or constitutional mechanism, the federal government clearly struggles with environmental policy.

The issue of national water policy bears consideration in this context because of the relatively recent establishment of the National Water Initiative (NWI) and Commission. This initiative, described in *Chapter Eight,* could be termed 'cooperative intergovernmental' for having its origins in agreement by the Council of Australian Governments (COAG), and because its national framework is overseen by a ministerial council but implemented by the states and territories. It remains to be seen, however, how the market-based approach embodied in the NWI squares the recognised need for restored environmental flows and improved ecological management in the face of constant pressures for increased consumption. In *Chapter Nine,* the federal role in adaptation again stems from a COAG agreement, but with the federal government largely providing funding for state and local projects, identified through localised planning. And in the area of disaster planning examined in *Chapter Ten,* it could be argued that the federal government's reluctance to cut emissions nationally is already having severe fiscal, ecological and human consequences at state and territory level. But the greatest reluctance to act by the federal government is certainly in population policy, with Australia's first population minister, Tony Burke, only appointed in 2010.

Towards policy success?

This book has been a study of policy inadequacy in a nation facing a suite of wicked environmental policy problems, captive to a 'quarry Australia' mindset, and with no clear federal leadership in terms of a pathway towards sustainability. It has argued that transformative, rather than palliative, policy solutions must be sought with the aim of achieving the principal element of policy success, which it has defined as ecological improvement and sustainability. In that respect, this book presents more than a study in policy process, programs and politics. It considers whether policy leads primarily to an improved environment, and it considers that question over time. Indeed, it emphasises the long term in its analysis of policy, but also in what is required of policy and its implementation and effects. This is because transformative policy cannot be short term if it is to achieve lasting ecological health. We have argued that, as the urgency to rescue nature increases and there is less time to act to save it, the policy benefits must be clear, transformative, durable and long term.

We have also stressed that Australia's environmental policy problems are longstanding and well known, which points to a failure to learn from them, to see them as significant, and in political terms to care much about the consequences.

Such failure is characteristic of a mindset that is blind to ecological concerns, an economy that counts the environment as a free resource, and policy that is ecologically dysfunctional. We see this not just as environmentally problematic but as a social issue too, because both the health and the actual survival of human society is threatened if ecosystems collapse, as we know all too well from the failures of past civilisations. It seems very likely that such lessons have not been learnt, that environmental decline will continue in Australia, that greenhouse gas emissions will soar, and that ecosystem health will decline, as will water availability and quality. Although we have discussed the policy and institutional complexities, we have also stressed, as many have before us, the need for political leadership, for strong advocacy, and for urgent action. Here we see an enhanced role for the state – indeed, for governments at all levels – to take on environmental concern, to integrate it thoroughly into decision-making, and to design new policies independent of political pressure and that will not cost the earth.

Notwithstanding its policy failure, Australia has some proud environmental traditions such as national parks and wilderness area declarations, with New South Wales Royal National Park the second oldest in the world, and more world heritage wilderness areas than any other nation. It is a country steeped in ecological awareness and an affinity with the natural environment that is much broader than its expression in the long tradition of green politics and the plethora of environmental movements to be found in Australia. In such a country there have been very many great environmental thinkers, scientists and analysts proposing, for example, that 'we have sufficient understanding today to solve the problems we face' but still questioning whether we have the wisdom and the courage to do so (Smith 1990, p. 170). Possibly one of the most useful texts to be read in search of ready access to improved policy is Lindenmayer *et al.*'s (2008) *Ten Commitments: Reshaping the Lucky Country's Environment*, in which forty experts propose ten solutions to each of over thirty Australian environmental problems.

This book has chartered the course of environmental policy failure, at times over the last century, but certainly to the extent that it can be seen to be steeped in Australian culture. Many negative trends date right back to European settlement in 1788. For environmental policy-making to become effective, we have argued, paradigms must change. As Milbraith (1996) suggested many years ago, we need to learn to think environmentally while there is still time. From environmental failure, the policy lesson learnt here is the need for the identification of Australia's environmental problems, and for the generation of sufficient knowledge, resources and political will to tackle them in a timely and effective fashion. We have stressed that policy must fundamentally transform: to embrace the principles of holism, connectivity and complexity; to acknowledge ecological thresholds; and to adopt precaution as standard practice. Australia must be recognised as atypically ecologically vulnerable. This should be planned for, as should the recognition and protection of the country's unique mega-biodiversity and the restructure of Australian industry away from exploitative natural resource dependency. At the same time, threatening processes must be recognised and reduced. Here, green politics has long played a key agenda-setting role that has led to Greens entering state parliaments and assemblies, and recently to the Australian Greens holding the balance of power in the Australian Senate. This may well be the political breakthrough that is needed to fundamentally transform Australian environmental policy.

Afterword

Policy futures?

As this book goes to press, the political settings of Australian public policy have changed abruptly from majoritarian to multi-party/member government with immediate implications for environmental policy. The federal Labor Government lost its House of Representatives majority at the 2010 federal election, whilst the balance of power in the Senate has now gone to the Australian Greens. The Greens' first lower House MP, plus three country-based Independent MPs, hold the balance of power in the lower house, while in the Senate nine Green Senators took office from July 2011 to hold the balance of power. Despite this seemingly precarious arrangement, the Government's legislative program has been proceeding though Parliament, albeit with many significant changes being made.

There has been immediate and unprecedented movement on climate change policy with the Australian Greens and Independents supporting the Labor Government in minority. Both major parties, Labor and Liberal, went to the 2010 federal election with no policy on climate change, with both advocating deferral till 2013 at the earliest. However, it is a condition of the Australian Greens supporting the Government that a carbon price be introduced. The elevation of climate policy as a top priority for action has triggered a flood of controversy, reminiscent of earlier controversy under the previous Rudd Labor Government, over the proposed super-profits tax on the mining industry. Once a carbon price (initially a tax but to be followed by an emissions trading scheme) is agreed by Labor, the Greens and Independents, the numbers ensure that it will proceed through Parliament.

This changed political and institutional setting, and the need for the Federal minority government to negotiate with either the Australian Greens or the Opposition in the Senate before legislation can pass, is the most significant breakthrough for the environment since Federation in 1901. This will be critical because many previous Labor policies, and indeed historical policies, are now so clearly failing–among them the recent Home Insulation Program which saw installers die in accidents. Environmental policy implementation failure has plagued Labor since it took government in 2007, with this Home Insulation

Program a classic case in point. It was part of a $3.9 billion Energy Efficient Homes stimulus plan that was subsequently found by the Auditor General to have involved more than 4,000 potential cases of fraud and 207 home fires. Most significantly, this failure has deterred energy efficiency action, which now languishes.

With the shift from majoritarian to multi-party/member government, for the first time since the stampede to neoclassical economic orthodoxy began decades ago new and radical perspectives on Australia's predicament are opening up. There is an unprecedented opportunity, whilst this precarious situation lasts, to elevate environmental concerns above the bi-partisan practice of 'putting the economy first'. There is now an opportunity to write the environment and sustainability into future versions of Australian economics, and in doing so to play to the expectations of many voters who so comprehensively dispensed with conservative government. Government has now been pulled markedly onto a more radical track than Labor had proposed in the 2010 election campaign, and the capacity for agenda setting on the environment is wide open.

It is too soon to speculate as to whether the voting trend delivering multi-party/member government will last in Australian federal politics even as far as the next election. Very often in Europe, when swinging away from social democratic governments such as Australia's Labor party, voters park their votes with independent or green parties rather than too readily embrace conservatism. Many in Australia are treating the swing to the Greens and the unusually large number of Independents elected as an aberration, a hiccup in the 'normal' pattern of two-party politics (or, more accurately, two-and-a-half party given that the conservative can only ever govern at the federal level in a coalition of Liberal and National Parties).

However, to do so requires commentators to ignore three important factors:

1. The growing Green vote, which has steadily increased over more than two decades, is now electing Members to the Australian Parliament as well as to state Houses besides just Tasmania where they have sat continuously since 1982 (Crowley 2008). In the face of a federal electoral system designed to disenfranchise smaller parties in its lower house, this is significant.

2. There is increasing popular disenchantment in Australia with the bi-partisan political orthodoxy that has dominated policy settings of 'small', stripped back government, low taxation, privatisation and unrestricted development–the latter invariably propped up by long-standing networks of subsidies and often at the heart of environmental despoliation.

3. The presence of Greens and Independents, with significant power, has focussed attention back onto policy issues which have been ignored and excluded by the prevailing neoconservative orthodoxy since the early 1980s in Australia. These issues include widespread concern over the

> environment, but they also include concern for human rights and humanitarian crises, and Australia's fall from grace as an honest international broker.

If the Parliamentary strength of the Greens and Independents survives the 2013 federal election, then it may well be safer to conclude that Australian politics is indeed undergoing a sea-change which should bode well for the environment. Is it hopelessly utopian, however, to assume that with changed politics will follow changed policy, and more targeted, better funded programs with implementation rigorous enough to redress the worst of Australia's environmental policy failure? The Greens are, after all, only minority supporters of government, not majority power holders in their own right, and this sort of fundamental policy change may yet be several generations hence.

KJ Walker and K Crowley
October 2011

References

AAP (Australian Associated Press) 2010a (26 July), 'Labor has no climate policy: Richo', *The Sydney Morning Herald*, <http://news.smh.com.au/breaking-news-national/labor-has-no-climate-policy-richo-20100726-10srp.html>.

AAP (Australian Associated Press) 2010b (15 September), 'Turnbull sticks to his guns on climate change', *The Sydney Morning Herald*, <http://www.smh.com.au/environment/climate-change/turnbull-sticks-to-his-guns-on-climate-change-20100915-15bjl.html>.

ABARE (Australian Bureau of Agricultural and Resources Economics) 2009, *Opportunities for Forestry under the Carbon Pollution Reduction Scheme: An Examination of Some Key Factors*, Commonwealth of Australia, Canberra.

ABARE (Australian Bureau of Agricultural and Resources Economics) 2010, *Australian Commodity Statistics 2009*, Australian Bureau of Agricultural Resources and Economics, Canberra.

ABC (Australian Broadcasting Corporation) 2004 (7 September), 'Leaked documents reveal fossil fuel influence in White Paper', *PM*, ABC Radio National, viewed 22 May 2011, <http://www.abc.net.au/pm/content/2004/s1194166.htm>.

ABC (Australian Broadcasting Corporation) 2009 (12 September), 'Ross Garnaut joins the 7.30 Report', *The 7.30 Report*, television program, ABC TV, 12 September 2009, transcript of interview available at <http://www.abc.net.au/7.30/content/2009/s2712070.htm>.

ABC Online 2006 (16 August), 'Commonwealth stays clear of carbon scheme', ABC, <www.abc.net.au/news/newsitems/200608/s1715997.htm>.

ABS (Australian Bureau of Statistics) 1992, *Striking a Balance: Australia's Development and Conservation*, Cat. no. 1314.0, ABS , Canberra.

ABS (Australian Bureau of Statistics) 2004, *Water Account Australia 2000-01, Cat.no.4610.0*, ABS, Canberra.

ABS (Australian Bureau of Statistics) 2006, *Agriculture Census*, ABS, Canberra.

ABS (Australian Bureau of Statistics) 2010, '3412.0 Migration, Australia (Replacement content)', ABS, Canberra.

Abetz, E, 2004, Adjournment: Nongovernment Organisations, *Hansard*, Australian Parliament, 15 April.

Abetz, E, 2006, 'Electoral reform: The Howard Government's agenda', address to the Sydney Institute, October 4, *The Sydney Papers*, vol. 18, viewed 5 November 2006, <wwww.thesydneyinstitute.com.au/publications/>.

ABS (see Australian Bureau of Statistics).

ACS (see Australian Coastal Society).

Adams, G & Hine, M 1999, 'Local environmental policy-making in Australia', in K Walker & K Crowley (eds), *Australian Environmental Policy 2: Studies in Decline and Devolution*, UNSW Press, Sydney, pp. 186-203.

Adams, J 2009, 'Risk management: The economics and morality of safety revisited', in C Dale & T Anderson (eds), *Safety-Critical Systems: Problems, Process and Practice,* Springer, New York.

AFSL (Australian Forestry Standard Ltd) 2007, *Forest Management-Economic, Social, Environmental and Cultural Requirements for Wood Production,* Australian Forestry Standard Ltd, Canberra.

Aleklett, K, Höök, M, Jakobsson, K, Lardelli, Snowden, S, & Söderbergh, B 2010, 'The peak of the oil age: analysing the world oil production reference scenario in World Energy Outlook 2008', *Energy Policy,* vol. 38, no. 3, pp. 1398–1414.

Alexander, K, 2010, personal interview.

Alexandra, J & Eyre, D 1993, 'Water and environment' in M Johnson & S Rix (eds), *Water in Australia, Managing Economic, Environmental and Community Reform,* Pluto Press, Leichhardt.

Alexandra, J & Riddington, C 2004, *Climate Change Impacts and Adaptation for Victoria,* Research Report, School of Social Science and Planning, RMIT University, Melbourne.

Allan, T 2004, 'Water resource development and the environment in the 20th century: first taking then putting back', in J Rodda & L Ubertini (eds), *The Basis of Civilization – Water Science?,* International Association of Hydrological Sciences.

Allison, I, Bindoff, NL, Bindschadler, RA, Cox, PM, de Noblet, N, England, MH, Francis, JE, Gruber, N, Haywood, AM, Karoly, DJ, Kaser, G, Le Quéré, C, Lenton, TM, Mann, ME, McNeil, BI, Pitman, AJ, Rahmstorf, S, Rignot, E, Schellnhuber, HJ, Schneider, SH, Sherwood, SC, Somerville, RCJ, Steffen, K, Steig, EJ, Visbeck, M, & Weaver, AJ 2009, *The Copenhagen Diagnosis: Updating the World on the Latest Climate Science,* Climate Change Research Institute, UNSW, Sydney, viewed 22 May 2011, <http://www.copenhagendiagnosis.org/default.html>.

Althaus, C, Bridgman, P & Davis, G 2007, *The Australian Policy Handbook,* 4th edn, Allen & Unwin, North Sydney.

ANCOLD (Australian National Committee on Large Dams Incorporated) 2010, <http://www.ancold.org.au/content.asp?PID=10005>.

Arndt, D Baringer, M & Johnson, M (eds) 2010, *State of the Climate in 2009,* National Oceanic and Atmospheric Administration (NOAA), USA.

ABS (Australian Bureau of Statistics) 2001, *Australia: State of the Environment 2001,* Department of the Environment and Heritage, CSIRO Publishing, Collingwood, Vic.

ABS (Australian Bureau of Statistics) 2006, *Australia: State of the Environment 2006,* published online by Department of the Environment, Sport and Territories, <http://www.environment.gov.au/soe/2006/publications/report/index.html>.

Attiwill, P 2010, 'Bushfires: managing the risk', *Geodate,* vol. 23, no. 1, pp. 6–8.

Audefroy, JF 2010, 'Post-disaster emergency and reconstruction experiences in Asia and Latin America: An assessment', *Development in Practice,* vol. 20, no. 6, pp. 664–677.

Auditor General (The) 2004, *The Administration of the National Action Plan for Salinity and Water Quality Audit,* Australian National Audit Office, Canberra.

Auditor General (The) 2008, *Regional Delivery Model for the Natural Heritage Trust and the National Action Plan for Salinity and Water Quality*, Australian National Audit Office, Canberra.

Auspoll 2008, *Climate Institute Climate Change Omnibus Report,* March, The Climate Institute, viewed 22 May 2011, <http://www.climateinstitute.org.au/index.php?option=com_content&view=category&layout=blog&id=131&Itemid=65>.

Austin, P 2010a, 'Premier's moment of truth', *The Age,* August 2, p. 1.

Austin, P 2010b, 'Opposition slams Brumby on bushfire protections', *The Age*, July 30, p. 2.

Australian Academy of Science 1994, Proceedings of the Symposium, *Population 2040: Australia's choice,* Australian Academy of Science, Canberra, <http://www.science.org.au/events/sats/sats1994/> and <http://www.science.org.au/events/sats/sats1994/Population2040-section8.pdf>.

Australian Academy of Science 2010, *The Science of Climate Change: Questions and Answers*, AAS, Canberra.

Australian Bureau of Statistics, Australian Bureau of Agricultural and Resource Economics & Bureau of Rural Sciences 2009, 'Socio-Economic Context for the Murray–Darling Basin', *Descriptive report, MDBA Technical Report Series*, Basin Plan, BP02, AGPS, Canberra.

ACS (Australian Coastal Society) 2010, *Policy on Coastal Planning and Infrastructure Reform*, July 2010, ACS, viewed 28 November 2010, <http://australiancoastal society.org/wp-content/uploads/ACS-Policy-on-Coastal-Planning-and-Infrastructure-Reform-July-2010.pdf>.

Australian Competition & Consumer Commission 2009, *Water Trading Rules: Draft Advice*, Australian Competition & Consumer Commission, Canberra.

Australian Government 1999, *Government Response to the Report by the House of Representatives Standing Committee for Long Term Strategies: Australia's Population Carrying Capacity: One Nation—Two Ecologies,* AGPS, Canberra.

Australian Academy of Science 2004, *Securing Australia's Energy Future,* Department of Prime Minister and Cabinet, Canberra.

Australian Academy of Science 2007, *2006-07 Natural Heritage Trust Annual Report*, Australian Government, Canberra.

Australian Academy of Science 2007, *Intergenerational Report 2007*, Australian Government/Treasury, viewed 27 December 2010, <http://www.treasury.gov.au/contentitem.asp?NavId=&ContentID=1239>.

Australian Academy of Science 2008, 'Emissions Trading in the Rest of the World', Fact Sheet.

Australian Academy of Science 2009, *Australia's Fourth National Report to the United Nations Convention on Biological Diversity*, <http://www.cbd.int/doc/world/au/au-nr-04-en.pdf?source=cmailer>.

Australian Academy of Science 2009, *National Energy Security Assessment*, Department of Resources, Energy & Tourism, Department of Climate Change, Canberra.

Australian Academy of Science 2010, *Intergenerational Report 2010, Australia to 2050: future challenges – the 2010 intergenerational report*, Australian Government/Treasury, viewed 27 December 2010, <http://www.treasury.gov.au/igr/igr2010/report/pdf/IGR_2010.pdf >.

Australian Academy of Science 2010, *Caring for Our Country Business Plan 2010–11*, Australian Government, Canberra.

Australian Government 2011, *Emissions Trading Schemes by Country*, Fact Sheet. Viewed 28 June 2011 <http://www.climatechange.gov.au/government/international/global-action-facts-and-fiction/ets-by-country.aspx>.

Australian State of the Environment Advisory Council 1996, *Australia: State of the Environment 1996*, AGPS/CSIRO Publishing, Collingwood, Vic. <http://www.environment.gov.au/soe/2006/publications/report/index.html>.

Bachelard, M 2010a, 'Feeling the heat: Labor will be ducking for cover over the Royal Commission findings', *The Age*, July 25.

Bachelard, M 2010b, 'Shire "backsliding" on bushfire policy', *The Age*, July 4, p. 7.

Balderstone, S, 2008, personal interview.

Ball, J, Donnelley, L, Erlanger, P, Evans, R, Kollmorgen, A, Neal, B, & Shirley, M 2001, 'Inland Waters', in *Australia, State of the Environment Report 2001 (theme report)*, CSIRO Publishing, Canberra.

Barnett, J & O'Neill, S 2010, 'Maladaptation', *Global Environmental Change*, vol. 20, pp. 211–213.

Barr, N 2008, 'The social landscapes of rural Victoria', in C Pettit W Cartwright, I Bishop, K Lowell, D Pullar & D Duncan (eds), *Landscape analysis and visualisation*, Springer, Berlin.

Bartlett, Albert A 2006, 'A depletion protocol for non-renewable natural resources: Australia as an example', *Natural Resources Research*, vol. 15, no. 3, pp. 151–164.

Baum, S, Horton, S, Low Choy D & Gleeson B 2010, *Climate Change, Health Impacts and Urban Adaptability: Case Study of Gold Coast City*, Research Monograph 11, Urban Research Program, Griffith University, Brisbane.

Beale, B 27 October 2010 'Trees not so large carbon sinks: study', University of New South Wales, Faculty of Science, viewed 28 October 2010, <http://www.science.unsw.edu.au/news/trees-not-so-large-carbon-sinks-study>.

Beck, U 1992, *Risk Society: Towards a New Modernity*, Sage, London.

Beder, S 1989, *Toxic Fish and Sewer Surfing: How Deceit and Collusion are Destroying our Great Beaches*, Allen & Unwin, North Sydney.

Beder, S 1997, *Global Spin*, Scribe, Melbourne.

Bell, SR, & Head, B (eds) 1994, *State, Economy, and Public Policy in Australia*, Oxford University Press, Melbourne.

Bell, SR & Quiggin, J 2008, 'The limits of markets: the politics of water management in rural Australia', *Environmental Politics*, vol. 17, no. 5, pp. 712–729.

Berger, C 2010, *Nomination Form For Adding a Threatening Process under the Environment Protection and Biodiversity Conservation Act 1999 (EPBC Act)*, 22 March 2010, on behalf of the Australian Conservation Foundation (ACF), on line at <http://www.acfonline.org.au/uploads/res/EPBC_nomination_22-3-10.pdf>.

Berkovic, N 2010, 'Green loans latest example of climate waste', *The Australian*, 9 July, viewed 4 July, <http://www.theaustralian.com.au/politics/green-loans-latest-example-of-climate-waste/story-e6frgczf-1225889581536>.

Berkovic, N 2009 (5 November), 'Labor solar scheme burns sugar industry', *The Australian*, viewed 22 May 2011, <http://www.theaustralian.com.au/business/news/labor-solar-scheme-burns-sugar-industry/story-e6frg90f-1225794498903>.

Betts, K 2010a, 'A bigger Australia: opinions for and against', *People and Place*, vol. 18, no. 2, pp. 25-38.

Betts, K 2010b, 'Attitudes to immigration and population growth in Australia 1954 to 2010: an overview', in *People and Place*, vol. 18, no. 3, pp. 32–51

Birch, C 1975, *Confronting the Future*, Penguin, Harmondsworth.

Birrell, R & Healy, E 2008, 'Labor's Greenhouse Aspirations', *People and Place*, vol. 16, no. 2, pp. 4, 11.

Birrell, R, Hill, D, & Stanley, J (eds) 1982, *Quarry Australia? Social and Environmental Perspectives on Managing the Nation's Resources*, Oxford University Press, Melbourne, chs 1-3.

Blackmore, D 2001 (10 January), 'Water, salinity and the politics of mutual obligation', in *The Politics of Water*, Alfred Deakin Lecture, broadcast on 20 May 2001, ABC Radio National, Australian Broadcasting Corporation, <http://www.abc.net.au/rn/deakin/content/session_12.html>.

Blackmore, D 2002, 'Protecting the Future', in D Connell (ed), *Uncharted Waters*, Murray–Darling Basin Commission, Canberra.

Blainey, G 1963, *The Rush That Never Ended*, Melbourne University Press, Carlton.

Boardman, R 1990, *Global Regimes & Nation States: Environmental Issues in Australian Politics*, Carleton University Press, Ottawa.

Boin, A 2009, 'The new world of crises and crisis management: Implications for policymaking and research', *Review of Policy Research*, vol. 26, no. 4, pp. 367–377.

Bolton, G 1992, *Spoils & Spoilers: A History of Australians Shaping their Environment*, 2nd edn, Allen & Unwin, North Sydney.

Bonnano, A & Constance, D 2001, 'Globalization, fordism, and post-fordism in agriculture and food: A critical review of the literature', *Culture and Agriculture*, vol. 23, no. 2, pp. 1–18.

Bonyhady, T 1993, *Places Worth Keeping: Conservationists, Politics & Law*, Allen & Unwin, North Sydney.

Bourke, W 2010, 'Population: It's the infrastructure, stupid. Or is it?', on-line article for ABC's *The Drum*, Australian Broadcasting Corporation, <http://www.abc.net.au/unleashed/39930.html>.

Bowman, DMJS, Balch JK, Artaxo, P, Bond, WJ, Carlson, JM, Cochrane, MA, D'Antonio, CM, DeFries, RS, Doyle, JC, Harrison, SP, Johnston, FH, Keeley, JE, Krawchuk, MA, Kull, CA, Marston, JB, Moritz, MA, Prentice, IC, Roos, CI, Scott, AC, Swetnam, TW, van der Werf, GR & Pyne, SJ 2009, 'Fire in the earth system', *Science*, vol. 324, pp. 481–484.

Boyden, S, Dovers, S, & Shirlow, M 1990, *Our Biosphere Under Threat*, Oxford University Press, Melbourne.

Bradsen, J 2000, 'Soil conservation: history, law and learning', in S Dovers (ed.), *Environmental History and Policy: Still Settling Australia*, Oxford University Press, South Melbourne.

Brissenden, M 2006, 'PM changes tack on climate change, wins on IR', *7:30 Report*, 14 November, television program, ABC 1, transcript, Australian Broadcasting Corporation, Sydney.

Bromby, R 1986, *Unlocking the Land: The Saga of Farming in Australia*, Lothian Publishing, Melbourne.

Brown, B 2001, media release, 20 July, viewed 22 May 2011, <http://bob-brown.greensmps.org.au/content/media-release/greenhouse-gas-abatement-program-corporate-welfare-instead-climate-care>.

Brown, H 1954, *The Challenge of Man's* Future, Viking Press, New York.

Brown, VA (ed.) 2010, *Tackling Wicked Problems*, Earthscan, London.

Brumby, J 2009, 'Justice Bernard Teague to head Fires Royal Commission', *Media Release*, Premier of Victoria, February 14, Melbourne.

Buckman, G & Diesendorf, M 2010, 'Design limitations in Australia renewable energy policies', *Energy Policy*, vol. 38, pp. 3365–76; addendum, vol. 38, Issue 7, pp. 7539–40.

Bulkeley, H & Betsill M 2005, 'Rethinking sustainable cities: Multilevel governance and the 'urban' politics of climate change', *Environmental Politics*, vol. 14 no. 1, pp. 42–63.

Bulkeley, H & Newell P 2010, *Governing Climate Change*, Routledge, London.

Bunker, R & Houston, P 2003, 'Prospects for the rural-urban fringe in Australia: Observations from a brief history of the landscapes around Sydney and Adelaide', *Australian Geographical Studies*, vol. 41, Issue. 3, pp. 233–247.

Burgin, S 2008, 'BioBanking: an environmental scientist's view of the role of biodiversity banking offsets in conservation', *Biodiversity Conservation*, vol. 17, pp. 807–816.

Burke, T 2010, 'Foreword by the Hon. Tony Burke, Minister for Sustainability, Environment, Water, Population and Communities', *A sustainable Population Strategy for Australia: Issues paper*, Commonwealth of Australia 2010, p. 1, <http://www.environment.gov.au/sustainability/population/publications/issues-paper.html>.

Burnley, IH, Fagan RH & Murphy, P 1997, *Immigration and Australian Cities*, Federation Press, Leichhardt, NSW.

Burns, A & Eltham, B 2010, 'Catastrophic failure theories and disaster journalism: Evaluating media explanations of Australia's "Black Saturday" bushfires', *Media Information Australia* (in press).

Burton, I 2010, 'Forensic disaster investigations in depth', *Environment*, vol. 52, no. 5, pp. 36–41.

Burton, D & Dredge, D 2007, 'Framing climate: Implications for local government policy response capacity', paper presented to the *State of Australian Cities* conference, UNSW, Sydney.

Butlin, NG, Barnard, A, & Pincus, JJ 1982, *Government and Capitalism: Public and Private Choice in Twentieth Century Australia,* Allen & Unwin, North Sydney.

Buxton, M 2010, 'Next generation needs action now', *The Age* (Melbourne), August 5, p. 17.

Buxton, M, Alvarez, A, Butt, A, Farrell, S, & O'Neill, D 2008, *Planning Sustainable Futures for Melbourne's Peri-urban Region*, RMIT University, Melbourne.

Buxton, M, Haynes, R, Mercer, D & Butt, A 2010, 'Vulnerability to bushfire risk at Melbourne's urban fringe: the failure of regulatory land use planning', *Geographical Research* (in press).

Buxton, M, Tieman, G, Bekessy, S, Budge, T, Coote, M, Mercer, D & Morcombe, J 2006, *Change and Continuity in Peri-urban Australia: State of the Regions*, School of Global Studies, Social Science and Planning, Melbourne.

Byrne, J, Gleeson, B, Howes, M & Steele, W 2009, 'Climate change & Australian urban resilience: The limits of ecological modernisation as an adaptive strategy', in S Davoudi, J Crawford & A Mehmood (eds), *Planning for Climate Change: Strategies for Mitigation and Adaptation*, Earthscan, London, Chapter 11, pp. 136–154.

Callaghan, A & Millington, T 1956, *The Wheat Industry in Australia*, Angus and Robertson, Melbourne.

Capling, A & Galligan, B 1992, *Beyond the Protective State: The Political Economy of Australia's Manufacturing Industry,* Cambridge University Press, Cambridge.

Carbon Positive 2009 (4 May), 'Australia delays cap and trade, lifts 2020 target', Carbon Positive, <http://www.carbonpositive.net/>.

Carden, MF 1999, 'Unsustainable Development in Queensland', Ch. 5 in KJ Walker & K Crowley, (eds), *Australian Environmental Policy 2: Studies in Decline & Devolution*, pp. 81–102, UNSW Press, Kensington.

Carey, A 1987, *Conspiracy or Groundswell?* pp. 3–19 in Coghill, K, (ed.).

Carr, B 2010, *Sustainable Development Panel Report*, Chair the Hon. Bob Carr, an Appendix to Burke 2010, <http://www.environment.gov.au/sustainability/population/publications/pubs/sustainable-development-panel-report.pdf>.

CAT (Centre of Alternative Technology) 2010, *Zero Carbon Britain 2030: A New Energy Strategy,* Centre of Alternative Technology, viewed 19 May 2011, <http://www.zerocarbonbritain.org>.

Cathcart, M 2009, *The Water Dreamers*, Text Publishing, Melbourne.

Cauchi, N 2011 (16 January), 'Wettest month in history for state's north and west', *The Age,* viewed 16 March 2011, <http://www.theage.com.au/environment/weather/wettest-month-in-history-for-states-north-and-west-20110115-19s16.html>

Chamie, J 2010, 'Is population growth a ponzi scheme?', *The Globalist*, 4 March 2010, <http://www.theglobalist.com/storyid.aspx?StoryId=8321>.

Christoff, P 2005, 'Policy autism or double-edged dismissiveness? Australia's climate policy under the Howard government', *Global Change, Peace and Security,* vol. 17, no. 1, pp. 29 – 44,

Christoff, P 2010, 'Touching the void: The Garnaut review in the chasm between climate science, economics and politics', *Global Environmental Change,* vol. 20, no. 2, pp. 214–217.

CIA (Central Intelligence Agency) 2011, 'Field listing: Population growth rate', *World Factbook,* viewed 4 June 2011, <https://www.cia.gov/library/publications/the-world-factbook/fields/2002.html>.

Cicin-Sain, B & Belifiore, S 2005, 'Linking Marine Protected Areas to integrated coastal and ocean management: a review of theory and practice', *Ocean and Coastal Management,* vol. 48, no. 11-12, 847–868.

Cicin-Sain, B & Knecht, R 1998, *Integrated Coastal and Ocean Management: Concepts and Practices,* Island Press, Washington, D.C.

CIE (Centre for International Economics) 2005, *Evaluation of the Natural Heritage Envirofund,* Centre for International Economics, Canberra.

Clark, S 1971, *The River Murray Question Part II,* Melbourne University Law Review, Melbourne.

Climate Institute (The) 2007, *Climate Institute: Marginal Electorates Exit Poll,* conducted by Australian Research Group Pty Ltd, November, viewed 22 May 2011, <http://www.climateinstitute.org.au/images/stories/exitpoll.pdf>.

Climate Institute (The) 2008, *Climate of the Nation: Australian attitudes to climate change and its solutions,* The Climate Institute, April, viewed 22 May 2011, <http://www.climateinstitute.org.au/index.php?option=com_content&view=category&layout=blog&id=131&Itemid=65>.

Climate Institute (The) 2009 (16 August), 'Politicians testing patience on climate change', The Climate Institute, <http://www.climateinstitute.org.au/component/content/article/112-blogs/504-politicians-testing-patience-on-climate-change>.

Clode, D 2010, *A Future in Flames,* Melbourne University Press, Carlton.

COAG 2002, *Natural Disasters in Australia: Reforming Mitigation, Relief and Recovery Arrangements,* A Report by a High Level Officials' Group, Canberra.

Cochrane, T 1989, *Blockade: the Queensland Loans Affair,* University of Queensland Press, St Lucia.

Coffey, B, Fitzsimons, JA & Gormly, R forthcoming, 'Strategic public land use assessment and planning in Victoria, Australia: Four decades of trailblazing but where to from here?', *Land Use Policy* (available at: <http://www.deakin.edu.au/dro/eserv/DU:30014919/fitzsimons-strategicpubliclanduse-2007.pdf>.

Coghill, K (ed.) 1987, *The New Right's Australian Fantasy,* McPhee Gribble, Fitzroy, Vic.

Cohen, B 2009, 'Abbott taps into climate of confusion' *The Australian,* 22 December, viewed 4 July, <http://www.theaustralian.com.au/news/opinion/abbott-taps-into-a-climate-of-confusion/story-e6frg6zo-1225812552415>.

Commonwealth of Australia 1992, *National forest policy statement: A New Focus for Australia's Forests,* Commonwealth of Australia, Canberra.

Commonwealth of Australia 1998a, *Australia's Ocean Policy: Volume 1*, Commonwealth of Australia, Canberra.

Commonwealth of Australia 1998b, *Australia's Ocean Policy: Volume 2*, Commonwealth of Australia, Canberra.

Commonwealth of Australia 2002, *Our vital resources: National Action Plan for Salinity and Water Quality*, Department of Environment, Heritage and the Arts, Viewed September 24, 2002, <http://www.napswq.gov.au/ publications/vital_resources.html>.

Commonwealth of Australia 2006, *Uranium Mining, Processing and Nuclear Energy Review: Opportunities for Australia?*, Report to the Prime Minister by the Uranium Mining, Processing and Nuclear Energy Review Taskforce, December 2006, Department of the Prime Minister & Cabinet, Canberra.

Commonwealth of Australia 2007, *Tackling Wicked Problems: A Public Policy Perspective,* Australian Public Service Commission, Canberra.

Conacher, A & Conacher, J 2000, *Environmental Planning and Management in Australia,* Oxford University Press, South Melbourne.

Congleton, RD 2006, 'The story of Katrina: New Orleans and the political economy of catastrophe', *Public Choice,* 127, pp. 5–30.

Connell, D 2007a, *Sustainability and the National Water Initiative,* Land and Water Australia, Canberra.

Connell, D 2007b, *Water Politics in the Murray–Darling Basin*, The Federation Press, Sydney.

Connell, RW 1977, Ruling Class, Ruling Culture, Cambridge University Press, Cambridge.

Connolly, WE 1969, 'The challenge to pluralist theory', in EE Connolly (ed), *The Bias of Pluralism*, Atherton Press, New York.

Connors, T 1970, 'Closer settlement schemes', *Australian Quarterly,* vol. 42, no.1, pp. 72–85.

Cork, S & Delaney, K 2005, *Thinking about the Future of Australia's Landscapes,* Land & Water Australia, Canberra.

Cornwell, A, Travis, J & Gunasekera, D 1997, *Framework for Greenhouse Emissions Trading in Australia*, Industry Commission Staff Research Paper, AGPS, Canberra.

Cottrell, A 2005, 'Communities and bushfire hazard in Australia: More questions than answers', *Environmental Hazards,* no. 6, no. 2, pp. 109–114.

Council of Australian Governments (COAG) 2007, *National Climate Change Adaptation Framework*, COAG, Canberra.

Council of Australian Governments 2004, *Intergovernmental Agreement on a National Water Initiative,* Available for download at <http://www.nwc.gov.au/resources/documents/Intergovernmental-Agreement-on-a-national-water-initiative.pdf>.

CRC (Cooperative Research Centre) 2011, website, Department of Innovation, Industry, Science & Research, viewed 22 May 2011, <https://www.crc.gov.au/Information/default.aspx>.

Crosby, AW 1986, *Ecological Imperialism*, Cambridge University Press, Cambridge.

Crough, G & Wheelwright, T 1982, *Australia: A Client State*, Penguin, Ringwood, Vic.

Crowley, K 2001, 'Effective environmental federalism?: Australia's Natural Heritage Trust', *Journal of Environmental Policy and Planning*, vol. 3, pp. 255-272.

Crowley, K 2007, 'Is Australia faking it? The Kyoto Protocol & the greenhouse policy challenge', *Global Environmental Politics*, vol. 7, no. 4, pp. 118–139.

Crowley, K 2008, 'The place of nature? Electoral politics and the Tasmanian Greens', *People and Place*, vol. 16, no. 2, pp. 12 - 21.

Crowley, K. 2010, 'Climate clever: Kyoto & Australia's decade of recalcitrance', in K Harrison & L McIntosh Sundstrom (eds), *Global Commons, Domestic Decisions: The Comparative Politics of Climate Change*, MIT Press, Cambridge, Massachusetts, pp. 201–228.

CSIRO & Land and Water Australia 2000, *Inter-State Water Trading; A Two Year Review: Executive Summary*, CSIRO Publishing, Canberra.

CSIRO 2008, *Water availability in the Murray Darling Basin, A report to the Australian Murray–Darling Basin Sustainable Yields Project*, CSIRO, Canberra.

CSIRO and Bureau of Meteorology 2007, *Climate Change in Australia: Technical Report 2007*, CSIRO, Canberra.

CSIRO and Bureau of Meteorology 2010, *State of the Climate*, CSIRO/BoM, Melbourne.

Cullen, P, 2004 (December), interview, interviewer D Connell, oral history collection Australian National Library, Canberra.

DAFF (Department of Agriculture Fisheries and Forestry) 2008, *The Changing Face of Australia's Forests: A Summary of Major Changes in Australia's Forests Since 1992*, Department of Agriculture, Fisheries and Forestry, Canberra.

Dahl, RA 1961, *Who Governs: Democracy and Power in an American City*, Yale University Press, New Haven.

Daley, J & Edis, T 2010, *Restructuring the Australian Economy to Emit Less Carbon*, Grattan Institute Report No. 2010–2 April 2010, The Grattan Institute, Melbourne, <http://www.grattan.edu.au>.

Daly, HE 2002, 'Sustainable development: Definitions, principles, policies', Speech to the World Bank, 30 April 2002, <http://www.earthrights.net/docs/daly.html>.

David, R 2007, 'The Indian middle class gradually changing its ways', *Forbes*, November 2007, <http://www.forbes.com/2007/11/10/india-middleclass-survey-face-markets-cx_rd_1108autofacescan01.html>.

Davidson BR 1965, *The Northern Myth: Limits to Agricultural and Pastoral Development in Tropical Australia*, Melbourne University Press, Carlton.

Davidson, BR 1969, *Australia Wet or Dry? The Physical and Economic Limits to the Expansion of Irrigation*, Melbourne University Press, Carlton.

Davidson, BR 1981, *European Farming in Australia*, Elsevier Scientific Publishing Company, New York.

Davies, P, Harris J, Hillman T & Walker K 2008, *Sustainable Rivers Audit: A report on the ecological health of rivers in the Murray–Darling Basin*, prepared by the Independent Sustainable Rivers Audit Group for the Murray–Darling Basin Ministerial Council.

Davis, B 1996 'Achieving sustainable development: Scientific uncertainty and policy innovation in Tasmanian regional development', *Australian Journal of Public Administration*, vol. 55, no. 4, 100–108.

Department of Climate Change (DCC) 2008, *Climate Pollution Reduction Scheme: Australia's Low Carbon Future: White Papers Volumes 1 and 2*, Commonwealth of Australia, DCC, Canberra.

Department of Climate Change (DCC) 2009, *Climate Change Risks to Australia's Coasts: A First Pass National Assessment*, Commonwealth of Australia, DCC, Canberra.

Department of Climate Change (DCC) 2009, *Overview of National Carbon Accounting Toolbox and Reforestation Reporting Requirements*, Commonwealth of Australia, DCC, Canberra.

Department of Climate Change (DCC) 2010, *Adapting to Climate Change in Australia: An Australian Government Position Paper*, Commonwealth of Australia, DCC, Canberra.

DCE (Department of Conservation and Environment Victoria) 1991, *Water Victoria – the Next 100 Years*, State of Victoria, Melbourne.

Dedekorkut, A, Mustelin, J, Howes, M & Byrne, J 2010, 'Tempering growth: Planning for the challenges of climate change and growth management in SEQ', *Australian Planner*, vol. 47 no. 3, pp. 203–215.

DEFRA (Department for Environment, Food and Rural Affairs) 2007, *UK Climate Change Programme: Annual Report to Parliament July 2007*, DEFRA (www.defra.gov.ak).

Dempsey, R (ed.) 1974, *The Politics of Finding Out: Environmental Problems in Australia*, Cheshire, Melbourne.

Department of Climate Change (DCC) 2009, *ClimateQ: Toward a Greener Queensland*, DERM, Brisbane.

Department of Climate Change (DCC) 2010, *What's Being Done*, Office of Climate Change, Department of Environment and Resource Management, Brisbane.

DEWHA (Department of Environment, Water, Heritage and the Arts) 2009, *A Framework for Determining Commonwealth Environmental Water Actions*, available for download at <http://www.environment.gov.au/water/policy-programs/cewh/pubs/cehw-framework-discussion-paper.pdf>.

DEWHA (Department of Environment, Water, Heritage and the Arts) 2010, *Securing our Water Future*, Australian Government, viewed 23 October 2010, <http://www.environment.gov.au/water/publications/action/pubs/securing-water-future.pdf>.

DIP (Department of Infrastructure and Planning) 2009, *South East Queensland Regional Plan 2009-31*, DIP, Brisbane.

DEWHA (Department of Environment, Water, Heritage and the Arts) 2010, *South East Queensland Infrastructure Plan and Program 2010-31*, DIP, Brisbane.

Dessai, S & Hulme, M 2004, 'Does climate adaptation policy need probabilities?' *Climate Policy*, vol. 4, no. 2, pp. 107–128.

Diamond, J 1991, *The Rise and Fall of the Third Chimpanzee*, Random House, London.

Diamond, J 1998, *Guns, Germs, and Steel*, Random House, London.

Diamond, J 2005, *Collapse: How Societies Choose to Fail or Succeed*, Allen Lane, London.

Dick Smith Population Links 2011, viewed 30 June 2011 <http://dicksmithpopulation.com/links/>.

Diesendorf, M 2005, *Can Nuclear Energy Reduce CO2 Emissions?* Viewed 20 July 2011, <http://www.sustainabilitycentre.com.au/CT_nukes_CO2.pdf>.

Diesendorf, M 2006, *Nuclear Power: Not Green, Clean or Cheap*, 16 June, On Line Opinion: Australia's e-journal of social and political debate, viewed 28 March 2011, <http://www.onlineopinion.com.au/view.asp?article=4581>.

Diesendorf, M 2007, *Greenhouse Solutions with Sustainable Energy*, UNSW Press, Kensington.

Diesendorf, M 2009, *Climate Action*, UNSW Press, Kensington.

Diesendorf, M 2009, 'Population growth negates efforts to halt greenhouse harm', *The Canberra Times*, 15 April 2009.

Diesendorf, M 2009, *Climate Action: A Campaign Manual for Greenhouse Solutions*, UNSW Press, Sydney.

Diesendorf, M 2010, 'Ambitious target does not quite measure up', *Ecos* no. 157, p. 30, Oct-Nov.

Diesendorf, M 2011a (in press), 'Are energy security and effective climate change policies contradictory objectives?' in J Symons & L Anceschi (eds), *Energy Security in the Era of Climate Change: The Asia-Pacific experience,* Palgrave Macmillan, Melbourne.

Diesendorf, M 2011b (in press), 'Can a carbon price cut greenhouse gas emissions?' *Australian Options*, Winter.

Diesendorf, M & Hamilton, C (eds) 1997, *Human Ecology, Human Economy*, Allen & Unwin, Sydney.

Diesendorf M, Lamb D, Mathews J, Burke N & Pearman G 2010, *Fuelling Future Passenger Vehicle Use in Australia,* Report of Jamison Group to NRMA Motoring & Services, NRMA, Sydney, February.

DIP (see Department of Infrastructure and Planning).

DNRE (Department of Natural Resources and Environment) 2001, *The Value of Water – A guide to Water Trading in Victoria*, State Government of Victoria, Melbourne.

Dodge, J 1999, 'Some bioregional theory and practice', in J Dryzek and D Schlosberg (eds), *Debating the Earth: The Environmental Reader*, Oxford University Press, Oxford, pp. 363–373.

Dovers, S 1994, *Australian Environmental History*, Oxford University Press, Melbourne.

Dovers, S 1997, 'Sustainability: demands on policy', *Journal of Public Policy*, vol. 16, no. 3, pp. 303–318.

Dovers, S 2000, *Environmental History and Policy: Still Settling Australia*, Oxford University Press, Melbourne.

Dovers, S 2008, 'Policy and Institution Reforms', in D Lindenmayer, S Dovers, Molly Harriss Olsen & S Morton (eds), *Ten Commitments: Reshaping the Lucky Country,* CSIRO Publishing, Collingwood, Victoria.

Dovers, S & Wild River, S (eds) 2003, *Managing Australia's Environment,* Federation Press, Annandale, NSW.

Dowling, J 2010, 'City to "grow" 134,000 homes on farmland, *The Age* (Melbourne), July 30, p. 10.

DRET (Department of Resources, Energy & Tourism) website, *Energy White Paper High Level Consultative Committee,* viewed 22 May 2011, <http://www.ret.gov.au/energy/facts/white_paper/sub_process/Pages/default.aspx>.

Dryzek, J 1987, *Rational Ecology: Environment and Political Economy,* Basil Blackwell, Oxford.

Dryzek, J 1997, *The Politics of the Earth,* Oxford University Press, New York.

Dryzek, J, Downes, D, Hunold, C, & Schlosberg, D, (with Hernes, H-K) 2003, *Green States and Social Movements,* Oxford University Press, Oxford.

Dryzek, JS, Hunold, C, & Schlosberg, D (with Downes, D & Hernes, H-K) 2002, 'Environmental transformation of the state: The United States, Norway, Germany and the United Kingdom', *Political Studies,* vol. 50, issue. 4, pp. 659–682.

DSE (Department of Sustainability, Water and Environment) 2006, *State Water Report 2004-05 – A Statement of Victorian Water Resources,* DSE, Melbourne.

DSE (Department of Sustainability, Water and Environment) 2007, *State Water Report 2005-06 – A Statement of Victorian Water Resources,* DSE, Melbourne.

DSE (Department of Sustainability, Water and Environment) 2008, *Monthly Water Report,* June, DSE, Melbourne.

DSE (Department of Sustainability, Water and Environment) 2009, *Water trade in northern Victoria 2007/08,* DSE, Melbourne.

DSE (Department of Sustainability, Water and Environment) 2010, Victorian Water Accounts 2008–2009, DSE, Melbourne.

Dunlap, TR 1997, 'Ecology and environmentalism in the Anglo settler colonies', Ch. 5 in T Griffiths, & L Robin (eds), *Ecology & Empire,* Melbourne University Press, Carlton.

Dunlevy, G 2010, 'Tony Burke queries balance in Murray–Darling Basin guide', *The Australian,* Sydney.

Dunlop, M & Brown, P 2008, *Implications of Climate Change for Australia's National Reserve System: A Preliminary Assessment, CSIRO Sustainable Ecosystems Report: A Report to the Department of Climate Change and the Department of the Environment, Water, Heritage and the Arts,* CSIRO, Canberra.

DWR (Department of Water Resources) 1986, *Water Law Review Discussion Paper, Report No.1,* Victorian Government Printing Office, North Melbourne.

DWR (Department of Water Resources) 1987a, *Security for Major Water Allocations, Issues Paper,* Government of Victoria, Melbourne.

DWR (Department of Water Resources) 1987b, *Security for Major Water Allocations, Background Paper,* Government of Victoria, Melbourne.

Dwyer, G & Esnouf, G 2008, 'Inquiries and reviews in Victoria: Key recommendations and outcomes 1939–2008', in: *Proceedings,* International Bushfire Research Conference and 15th Annual AFAC Conference, Adelaide.

Eckersley, R 2004, *The Green State: Rethinking Democracy and Sovereignty*, MIT Press, Massachusetts.

Edwards, G 2003, 'Policy forum: water pricing and availability, water policy, setting the scene', *The Australian Economic Review*, vol. 36, no. 2, pp. 193–202.

Ellis, S, Kanowski, P, & Whelan, RJ 2004, *National Inquiry on Bushfire Mitigation and Management*, Council of Australian Governments, Canberra.

Emerson, C 2008, personal interview.

Energy Bulletin 2010, 'Are cities sustainable in a post-peak oil world?', (staff article surveying the debate), published February 1 2010, <http://www.energybulletin.net/node/51386>.

England, P 2007, *Climate Change: What are local governments liable for?* Issue Paper 6, Urban Research Program, Griffith University, Brisbane.

EREC (European Renewable Energy Council) 2010, *Rethinking 2050: A 100% renewable energy vision for the European Union,* European Renewable Energy Council, Brussels.

Eriksen, C & Gill, N 2010, 'Bushfire and everyday life: Examining the awareness-action "gap" in changing rural landscapes', *Geoforum* (in press).

Evans, N, Morris, C & Winter M 2002, 'Conceptualizing agriculture: a critique of post-productivism as the new orthodoxy', *Progress in Human Geography,* vol. 26, no.3, pp. 313–332.

Ferguson, S 2009, 'Malcolm and the malcontents', *Four Corners,* television program, ABC TV, transcript, 19 November, Australian Broadcasting Corporation, Sydney.

Fisher, R 2010, personal interview.

Fitzsimons, J, Legge, S, Traill, B & Woinarski, J 2010, *Into oblivion? The Disappearing Native Mammals of Northern Australia,* Nature Conservancy, Melbourne.

Flannery, T F 1994a, *The Future Eaters: An Ecological History of the Australasian Lands and People,* Reed, Chatswood, NSW.

Flannery, T, 1994b, 'Biological considerations in determining an optimum human population for Australia', Proceedings of the Symposium, *Population 2040: Australia's Choice,* Australian Academy of Science.

Foran, B & Poldy, F 2002a, *Future Dilemmas: Options to 2050 for Australia's population, technology, resources and environment,* online report for CSIRO and Department of Immigration and Multicultural and Indigenous Affairs, <http://www.cse.csiro.au/research/futuredilemmas/>.

Foran, B & Poldy, F 2002b, *Dilemmas Distilled: A Summary and Guide to the CSIRO Technical Report,* online report for CSIRO and Department of Immigration and Multicultural and Indigenous Affairs, <http://www.cse.csiro.au/research/futuredilemmas/>.

Franklin, R 2009, *Inferno: The Day Victoria Burned,* Slattery Media Group, Melbourne.

Frawley, K 1994, 'Evolving visions: Environmental management and nature conservation in Australia', Ch. 4 in S Dovers (ed.), *Australian Environmental History,* pp. 55–78.

Friedmann, H & McMichael P 1989, 'Agriculture and the state system: The rise and decline of national agricultures, 1870 to the present', *Sociologia Ruralis*, vol. 29, no. 2, pp. 93–117.

Fullerton, T 2002, 'Search for a Supermodel (the saga of *Future Dilemmas*)', *Four Corners*, television program, Australian Broadcasting Corporation, 5 November 2002, transcript at <http://www.abc.net.au/4corners/stories/s718523.htm>.

Fyfe, M 2010, 'State blamed for bushfires', *The Age* (Melbourne), October 3, p. 4.

Gans, J & Quiggin, J 2007 'The practicalities of emissions trading', *The Melbourne Review*, vol. 3, no. 2, November, pp. 60–65.

Gardner, A & Bowmer, K 2007, 'Environmental water allocations and their governance' in K Hussey & S Dovers (eds), *Managing Water for Australia: The Social and Institutional Challenges*, CSIRO Publishing, Collingwood.

Garnaut, R 2008, *Garnaut Climate Change Review Final Report*, Cambridge University Press, Cambridge.

Garnaut, R 2008b, 'Oiling the Squeeks' *Sydney Morning Herald*, 20 December, viewed 4 July 2011, <http://www.smh.com.au/news/environment/global-warming/oiling-the-squeaks/2008/12/19/1229189886229.html>.

Garnaut, R 2009 (9 March), *Four Corners*, television program, ABC 1, Australian Broadcasting Corporation, <http://www.abc.net.au/news/stories/2009/03/09/2510417.htm>.

Garnaut, R 2010, *Climate Change, China Booms and Australia's Governance Struggle in a Changing World*, 2010 Hamer Oration, The University of Melbourne and Hamer Family Fund, August 5.

GCCC (Gold Coast City Council) 2006, *Gold Coast City Community Profile*, GCCC, Southport.

GCCC (Gold Coast City Council) 2009, *Climate Change Strategy 2009–14*, DCC & GCCC, Southport.

Gentle, G & Olszak, C 2007, 'Water planning: principles, practices and evaluation', in K Hussey & S. Dovers (eds), *Managing Water for Australia: The Social and Institutional Challenges*, CSIRO Publishing, Collingwood.

Gettler, L 2010 (8 July), 'Putting a Price on Carbon', *The Sydney Morning Herald*, <http://www.smh.com.au/business/putting-a-price-on-carbon-20100707-100r3.html>.

Gilpin, Z 1980, *The Australian Environment: Twelve Controversial Issues*, Sun Books, Melbourne.

Gittins, R 2010 (20 December), 'Beware gurus selling high immigration', *Sydney Morning Herald*, <http://www.smh.com.au/business/beware-gurus-selling-high-migration-20101219-19201.html>.

Glazebrook, G, Rickwood, P, & Searle, G 2007, 'Urban structure and energy—a review', *Urban Policy and Research*, 2007, pp. 1-25, <http://www.fbe.unsw.edu.au/cf/staff/peter.rickwood/files/rgs_upr08_extract.pdf>

Goodchild, P 2010, *The coming chaos: oil, electricity, agriculture, population, tribulation, survival*, (abridged version) online book <https://docs.google.com/fileview?id=0B5F-idWfw7TeY2U4ZjE3ZmUtYjU5Ny00ZGQ5LTg4YTQtMGVmZjg5ZjQxZGI4&hl=en&authkey=CMDYyYcF>.

Grafton, RQ & Hussey, K 2007, 'Buying back the Living Murray: at what price?' *Australasian Journal of Environmental Management*, vol. 14, no. 2, pp. 74–81.

Grafton, RQ & Jiang, Q 2010, 'Economics of drought, water diversions, water recovery and climate change in the Murray–Darling Basin', *Centre for Water Economics, Environment and Policy (CWEEP) Research Paper 10–01*, available for download at <http://cweep.anu.edu.au/pdf/publications/research_papers/10-01_WaterEconomics.pdf>.

Grant, W 1985, 'Introduction', in W. Grant (ed), *The Political Economy of Corporatism*, St. Martin's Press, New York.

Grattan, M 2009, 'We will have climate policy', *The Age*, 2 September 2009.

Grattan, Michelle & Arup, Tony 2008 (17 December), 'Emissions goals split all comers', *The Age*, <http://www.theage.com.au/environment/emissions-goals-split-all-comers-20081216-6zuq.html>.

Gray, D 2010, 'Pressure to triple state fuel burn-off', *The Age*, August 4, p. 10.

Greenpeace Australia 2010, *Future Risk: The Increased Risk of Catastrophic Bushfires Due to Climate Change*, Report for Fire Brigade Union of New South Wales, Sydney.

Greens 2011, Australian Greens' Population Policy, viewed 30 June 2011, <http://greens.org.au/policies/environment/population>.

Griffiths, T 2001, *Forests of Ash: An Environmental History*, Cambridge University Press, Cambridge.

Griffiths, T 2009, *Submission to the 2009 Victorian Bushfires Royal Commission*, Submission No. 002.038.0084_R, 16 May.

Habermas, J, 1962, *The Structural Transformation of the Public Sphere: An Inquiry into a Category of Bourgeois Society*, Polity, Cambridge.

Habermas, J, 1981, *The Theory of Communicative Action*, Polity, Cambridge.

Hajkowicz, S 2009, 'The evolution of Australia's natural resource management programs: Towards improved targeting and evaluation of investments', *Land Use Policy*, vol. 26, no. 2, pp. 471–478.

Hall, C M 1992, *Wasteland to World Heritage: Preserving Australia's Wilderness*, Melbourne University Press, Carlton.

Hamilton, C 2001, *Running from the Storm: the Development of Climate Change Policy in Australia*, UNSW Press, Kensington.

Hamilton, C 2007, *Scorcher: The Dirty Politics of Climate Change*, Black Inc. Agenda, Melbourne, pp. 10, 14.

Hamilton, C & Maddison, S, 2007, *Silencing Dissent: How the Australian Government is Controlling Public Opinion and Stifling Debate*, Allen & Unwin, Crows Nest.

Hamilton, C & Throsby, D 1997, *The ESD Process: Evaluating a Policy Experiment*, Academy of Social Sciences, Canberra.

Hamstead, M, Bladwin, C & O'Keefe, V 2008, *Water Allocation Planning in Australia – Current Practices and Lessons Learned*, National Water Commission, Canberra.

Hancock, WH 1930, *Australia*, Ernest Benn Ltd, Melbourne.

Hansson, A & Bryngelsson M 2009, 'Expert opinions on carbon dioxide capture and storage—A framing of uncertainties and possibilities', *Energy Policy*, vol. 37, no. 6, pp. 2273–82.

Hardin, G 1968, 'The tragedy of the commons', *Science*, vol. 163, pp. 1243-1248.

Hardin, G 1972, *Exploring New Ethics for Survival: The Voyage of the Spaceship Beagle*, Viking Press, New York.

Harris, M 1979, *Cultural Materialism*, Random House, New York.

Harrison, K & McIntosh Sundstrom, L (eds) 2010, *Global Commons, Domestic Decisions: The Comparative Politics of Climate Change*, MIT Press, Cambridge, Massachusetts.

Harvey, N & Caton, B 2003, *Coastal Management in Australia*, Oxford University Press, Melbourne.

Haward, M & Vince, J 2008, *Oceans Governance in the Twenty-first Century: Managing the Blue Planet*, Edward Elgar, Cheltenham.

Hawke, A 2009, *The Australian Environment Act: Report of the Independent Review of the EPBC Act 1999*, Department of Environment, Water, Heritage and the Arts, Canberra.

Haynes, K, Handmer, J, McAneney, J, Tibbits, A & Coates, L 2010, 'Australian bushfire fatalities 1900-2008: exploring trends in relation to the *Prepare, stay and defend or leave early* policy', *Environmental Science & Policy*, vol. 13, no. 3, pp. 185–194.

Head, B 2008, 'Wicked problems in public policy', *Public Policy*, vol. 3, no. 2, pp. 101–118.

Head, B 2009, 'Adaptation to climate change: the evolution of policy and program ideas in Australia', in J. Martin, M. Rogers & C. Winter, (eds), *Climate Change in Regional Australia: social learning and adaptation*, Victorian Universities Regional Research Network Press, Ballarat, pp.14–29.

Henry, K 2009, 'The shape of things to come: Long run forces affecting the Australian economy in coming decades', address by Dr Ken Henry to the Queensland University of Technology Business Leaders' Forum, 22 October <http://www.treasury.gov.au/documents/1643/HTML/docshell.asp?URL=QUT_Address.htm>.

Herbohn, K, Harrison, S & Herbohn J 2000, 'Reforestation progress and forestry incentive schemes in Australia', in S. Harrison, J. Herbohn K. Herbohn (eds), *Sustainable Small-scale Forestry: Socio-economic Analysis and Policy*, Edward Elgar, Cheltenham, pp. 214–223.

Hertin, J, Turpenny J & Jordan, A 2009, 'Rationalising the policy mess? Ex ante policy assessment and the utilisation of knowledge in the policy process', *Environment & Planning A*, vol. 41, no. 5, pp. 1185–1200.

Horridge, M, Madden, J & Wittwer, G 2005, 'The impact of the 2002–2003 drought on Australia', *Journal of Policy Modelling*, vol. 27, no. 3, pp. 285–303.

HORSCCCWEA (House of Representatives Standing Committee on Climate Change, Water, Environment and the Arts) 2009, *Managing Our Coastal Zone in a Changing Climate: The Time to Act is Now*, Commonwealth of Australia, Canberra.

Howes, M 2005, *Politics and the Environment: Risk and the Role of Government and Industry*, Allen & Unwin, North Sydney.

Howes, M 2008, 'Rethinking Governance: Lessons in collaboration from environmental policy', *Australian Political Studies Association Annual Conference*, 6–9 July, University of Queensland, Brisbane.

Hudson, M, 1995, *Managing Without Profit: The Art of Managing Third-Sector Organisations,* Penguin Books, London.

Hughes, L 2003, 'Climate change and Australia: Trends, projections and impacts', *Austral Ecology,* vol. 28, pp. 423–443.

Huxley, J 2007, 'The great desert dream', *The Sydney Morning Herald,* viewed 23 October 2010,

IEA (International Energy Agency) 2010, *World Energy Outlook 2010,* executive summary, International Energy Agency, see p.6 and graph 'Oil production becomes less crude', viewed 19 May 2011, <http://www.worldenergyoutlook.org>.

Ingamells, P 2010, 'More to learn about fire before setting a burn-off target', *The Age* (Melbourne), July 30, p.10.

Ingram, JC, Franco, G, Rumbaitis-del Rio, C & Khazai, B 2006, 'Post-disaster recovery dilemmas: challenges in balancing short-term and long-term needs for vulnerability reduction', *Environmental Science & Policy,* vol. 9, Issue. 7-8, pp. 607–613.

IPPC (Intergovernmental Panel on Climate Change 2007, *IPCC Fourth Assessment Report,* viewed 22 May 2011, <http://www.ipcc.ch/publications_and_data/ar4/syr/en/contents.html>.

Ison, R & Collins, K 2008, 'Public policy that does the right thing rather than the wrong thing righter', unpublished paper presented to the *Analysing Collaborative and Deliberative Forms of Governance Workshop* 14 November Australian National University, Canberra.

Jacobson, MZ, & Delucchi, MA 2009, 'A path to sustainable energy by 2030', *Scientific American,* vol. 301, no. 5, November 2009, pp. 38–45, p. 39.

Jaensch, D 1997, *The Politics of Australia,* Allen & Unwin, North Sydney.

Janicke, M & Weidner, H 1995, *Successful Environmental Policy: A Critical Evaluation of 24 Countries,* Sigma, Berlin.

Johnson, M 1993 'The water industry overseas: Lessons for Australia' in M Johnson & S Rix (eds), *Water in Australia, Managing Economic, Environmental and Community Reform,* Pluto Press, Leichhardt.

Johnston, W 1988, *A Documentary History of Queensland,* University of Queensland Press, St Lucia.

Jones, S 2010, 'The national renewable energy target: An example of post-Machiavellian policy-making?' *Australian Journal of Public Administration,* vol. 69, no. 2, pp. 165–177.

Jones, B 1994, Speech by the Rt. Hon. Barry Jones, Member for Labor, presenting to the Australian House of Representatives the report *One Nation: Two Ecologies. Hansard,* 5 December 1994, p. 3915.

Kane, S & Shogren, J 2000, 'Linking adaptation and mitigation in climate change policy', *Climatic Change,* vol. 45, no. 1, pp. 75–102.

Katzenstein, PJ 1985, *Small States in World Markets: Industry Policy in Europe,* Cornell University Press, Ithaca.

Keane, B 2009, *Who Knew an Energy White Paper was Under Development?*, crikey.com, 4 March, viewed 19 May 2011, <http://www.crikey.com.au/2009/03/04/who-knew-an-energy-white-paper-was-under-development>.

Kelly, J 2011, 'Global warming will cause further extreme weather patterns, Climate Chief says' *The Australian* viewed 25 January 2011, <http://www.theaustalian.com.au/national-affairs/climate/global-warming-will-cause-further-extreme-weather-patterns--climate-change-chief-says/story-e6frg6xf-1225985772600>.

Kelly, P (16 July) 2008, 'Government caught in carbon crunch' *The Australian,* p. 14, viewed 4 July 2011, <http://www.theaustralian.com.au/national-affairs/garnaut-hits-at-failure-on-climate-change-policy/story-fn59niix-1225901841323>.

Kelly, P 2010, 'Garnaut hits failure on climate change policy', *The Australian*, 6 August.

Kellow, A 1992, *Saline solutions: Policy Dynamics in the Murray Darling Basin,* Geelong, Centre for Applied Social Research, Deakin University, Geelong.

King, D 2006, Statement by Sir David King to a British Parliamentary enquiry, July 2006. See <www.optimumpopulation.org/opt.release13Mar07.htm>.

Kissane, K 2010, *Worst of Days. Inside the Black Saturday Firestorm,* Hachette, Sydney.

Kraft, ME & Furlong, SR 2007, *Public Policy: Politics, Analysis and Alternatives,* CQ Press, Washington.

Lagadec, P 2009, 'A new cosmology of risks and crises: Time for a radical shift in paradigm and practice', *Review of Policy Research,* vol. 26, no. 4, pp. 473–486.

Lake, M 1987, *The Limits of Hope: Soldier Settlement in Victoria 1915-1938,* Oxford University Press, Melbourne.

Lake, P & Bond, N 2007, 'Australian futures: Freshwater ecosystems and human water usage', in *Futures,* vol. 39, no. 2–3, pp. 288–305.

Lambert, J 2008, personal interview.

Langsam, D 2010, 'Climate change elephant in room', *The Age,* August 5, p. 16.

Lardelli, M 2009, 'The population problem', *On Line Opinion,* 6 March 2009, <http://www.onlineopinion.com.au/view.asp?article=8635&page=0 >.

Lardelli, M 2010, 'Can we trust the Greens on population?', *On Line Opinion,* 20 August 2010, <http://www.onlineopinion.com.au/view.asp?article=8635&page=0>.

Latz, P 2007, *The Flaming Desert: Arid Australia – A Fire-Shaped Landscape,* P Latz, Alice Springs.

Lauder, S 2010, 'More extremes for Australia as climate change tightens its grip', *The World Today, ABC Radio National* 3 December 2010, viewed 14 December 2010, <http://www.abc.net.au/worldtoday/content/2010/s3083916.htm>.

Laver, M 1981, *The Politics of Private Desires,* Penguin, Harmondsworth.

Lazarow, N, Souter, R Fearon, R and Dovers, S (eds) 2006, *Coastal management in Australia: Key institutional and governance issues for coastal natural resource management and planning,* CRC for Coastal Zone, Estuary and Waterway Management, Brisbane.

Le Grand, C. (2010) 'Relief efforts shift to economic recovery', *The Australian,* August 14, p. 9.

Lee, J, 2 July 2010, 'How a $7m advertising campaign saved a fortune,' *The Sydney Morning Herald*, viewed 28 March 2011, <http://www.smh.com.au/business/how-a-7m-advertising-campaign-saved-a-fortune-20100701-zqtq.html>.

LETDF (Low Emissions Technology Demonstration Fund) webpage, Department of Resources, Energy and Tourism, viewed 22 May 2011, <http://www.ret.gov.au/energy/clean/cei/low_emissions_technology_demonstration_fund/Pages/LowEmissionsTechnologyDemonstrationFund.aspx>.

Liebrecht, T & Howes, M 2006, 'Collaboration: a solution to inter-jurisdictional strife?', *Governments and Communities in Partnership Conference*, Centre for Public Policy, 25–27 September, University of Melbourne.

Lindblom, C 1977, *Politics and Markets: The World's Political Economic Systems*, Basic Books, New York.

Lindenmayer, D, 2007, *On Borrowed Time: Australia's Environmental Crisis and What We Must Do About It*, Penguin Group and CSIRO Publishing, Camberwell, Victoria.

Lindenmayer, D, Dovers, S, Harriss Olson, M and S Morton 2008, *Ten Commitments: Reshaping the Lucky Country's Environment*, CSIRO Publishing, Collingwood, Vic.

Lines, WJ 1991, *Taming the Great South Land: a History of the Conquest of Nature in Australia*, Allen & Unwin, North Sydney.

Lines, W, & O'Connor, M 2010, *Overloading Australia: How Governments and Media Dither and Deny on Population*, 4th edn, Envirobook, Sydney.

Lockie, S 1993, *Landcare: Self-Help, Local Knowledge and the State*, Ecopolitics VII conference, Brisbane.

Lockie, S & Higgins, V 2007, 'Roll-out neoliberalism and hybrid practices of regulation in Australian agri-environmental governance', *Journal of Rural Studies*, vol. 23, Issue. 1, pp. 1–11.

Lockie, S & Vanclay, F 2000, 'Critical Landcare: Introduction' in S Lockie & F Vanclay (eds) *Critical Landcare*, Centre for Rural Social Research, Wagga Wagga.

Lohrey, A 2002, 'Groundswell: The rise of the Greens', *Quarterly Essay*, vol 8, p. 1–86.

Lowe, I 1991, 'Towards a Green Tasmania: Developing the "Greenprint"', *Habitat Australia*, August, pp. 12–14.

Lowe, I 1994, 'The greenhouse effect and the politics of long-term Issues', Chapter 13, in SR Bell & B Head (eds), *State, Economy, and Public Policy in Australia*, Oxford University Press, Melbourne.

Lowe, I 2005, *A Big Fix: Radical Solutions for Australia's Environmental Crisis*, Black Inc. Book, Melbourne.

Lucas, C 2009, *Climate Change Impacts on Fire Weather*, Centre for Australian Weather and Climate Research, Melbourne.

Lukes, S 1974, *Power: A Radical View*, Macmillan, London.

Lunn, S 2010, 'Premier weak on call for burns', *The Australian*, August 3, p. 4.

Lyster, R 2004, 'Common But Differentiated? Australia's response to global climate change', *Georgetown International Environmental Law Review*, vol. 16, no. 4, pp. 561–591.

Macpherson, CB 1962, *The Political Theory of Possessive Individualism: Hobbes to Locke*, Clarendon Press, Oxford.

Maddison, S & Denniss, R, 2005, 'Democratic constraint and embrace: Implications for progressive non-government advocacy organisations in Australia', *Australian Journal of Political Science,* vol. 40, no. 3, pp. 373–389.

Maddison, S & Edgar, G 2008, 'Into the lion's den: Challenges for the not-for-profits in their relationships with government', in J Barraket (ed), *Strategic Issues for the Not-for-Profit Sector*, UNSW Press, Sydney, pp. 188-211.

Makhijani, A 2007, *Carbon-Free and Nuclear-Free: A roadmap for U.S. Energy Policy,* RDR Books, Muskegon MI & IEER Press, Takoma Park MD.

Mann, M 1981, 'Capitalism and Militarism', Ch. 1 in M Shaw (ed), *War, State & Society*, Macmillan, London.

Manne, R 2009, 'Why we weren't warned: The Victorian bushfires and the Royal Commission', *The Monthly,* July, pp. 22–35.

Manning, P & Darby A 2010 (10 September), 'Timber giant concedes defeat in decades-old logging war', *The Age,* viewed 23 October 2010, <http://www.theage.com.au/national/timber-giant-concedes-defeat-in-decadesold-logging-war-20100910-153fm.html>.

Manning, P 2004, 'Keeping democracy in its place', in M. Kingston (ed), *Not Happy, John!*, Penguin Books, Melbourne.

Manning, P 2010, 'Climate another blow in a bad era for capitalism', *The Age,* February, p. 21.

Mansergh, I, Cheal, D & Fitzsimons, J 2008, 'Future landscapes in south-eastern Australia: the role of protected areas and biolinks in adaptation to climate change', *Biodiversity* vol. 9 no. 3, pp. 59–70.

Marr, A 2008, personal interview.

Marsh, D & McConnell, A 2010, 'Towards a framework for establishing policy success', *Public Administration: An International Quarterly,* vol. 88, no. 2, pp. 564-583.

Marsh, I (ed.) 1991, *The Environmental Challenge*, Longman Cheshire, Sydney.

Marshall, AJ 1966, *The Great Extermination*, Heinemann, London.

Mason, B 2004a, Adjournment: Non-government organisations, *Hansard,* Australian Parliament, 24 June, Canberra.

Mason, B 2004b, Taxation: Charitable institutions, *Hansard,* Australian Parliament, 1 March, Canberra.

Mason, IG, Page, SC & Williamson, AG 2010, 'A 100% renewable electricity generation system for New Zealand utilising hydro, wind, geothermal and biomass resources', *Energy Policy*, vol. 38, no. 8, pp. 3973–84.

McCutcheon, P 2009 (8 December), 'Renewable energy on the brink', *7.30 Report*, television program, Australian Broadcasting Corporation (ABC), viewed 22 May 2011, <http://www.abc.net.au/7.30/content/2009/s2765610.htm>.

McKee, JK 2009, 'Contemporary mass extinction and the human population imperative', *Journal of Cosmology*, 2009, vol. 2, pp. 300–308, <http://journalofcosmology.com/Extinction104.html>.

McKibben, WE, 2010, 'Breaking the growth habit', *Scientific American*, April 2010, <http://www.scientificamerican.com/article.cfm?id=breaking-the-growth-habit>.

McKinsey & Company 2009, *Pathways to a Low-Carbon Economy*, Version 2 of the global greenhouse gas abatement cost curve, viewed 22 May 2011, <http://www.mckinsey.com/en/Client_Service/Sustainability/Latest_thinking/Pathways_to_a_low_carbon_economy.aspx>.

McNamara, A 2009, 'The problem is us', speech to the Queensland Conservation Council Brisbane, 17 July 2009, citing World Bank figures. Note there are many such tables with minor differences—readily compared at <http://en.wikipedia.org/wiki/List_of_countries_by_GDP_(nominal)_per_capita#cite_note-1>.

McNamara, A 2010-2011, 'Why the city we build will never be the city we need', *Dissent* no. 34, Summer 2010–2011, pp. 24–30

MDBA (Murray–Darling Basin Authority) 2010, *Guide to the Proposed Basin Plan*, Murray–Darling Basin Authority, Canberra, <http://thebasinplan.mdba.gov.au/guide/>.

MDBC (Murray–Darling Basin Commission) 2005, *The Living Murray Business Plan*, Murray–Darling Basin Commission, Canberra.

MDBMC (Murray–Darling Basin Ministerial Council) 2002, *The Living Murray: A Discussion Paper on Restoring the Health of the River Murray, Stage 1: Informing and Engaging the Community*, Murray–Darling Basin Commission, Canberra.

Mees, P 2009, *Transport for Suburbia: Beyond the Automobile Age*, Earthscan, London and Sterling, VA.

Melbourne Water 2005, *Melbourne's Water Supply System*, Melbourne Water, Melbourne.

Menadue, P 2010, 'Who killed Kevin Rudd's prime ministership?' *ABC Drum Unleashed*, 28 June, Australian Broadcasting Corporation, Sydney, <http://www.abc.net.au/unleashed>.

Mendham, E & Curtis A 2010, 'Taking over the reins: Trends and impacts of changes in rural property ownership', *Society and Natural Resources*, vol. 23, no. 7, pp. 653–668.

Mercer, D 2000, *'A Question of Balance': Natural Resources Conflict Issues in Australia*, 2nd edn, Federation Press, Sydney.

Mercer D & Marden, P 2006, 'Ecologically sustainable development in a "Quarry" Economy: One step forward, two steps back', *Geographical Research*, June, vol. 44, no. 2, pp. 183–203.

Metz, B, Davidson, OR, Bosch, PR, Dave, R & Meyer, LA (eds) 2007, *Climate Change 2007: Mitigation of Climate Change*, Contribution of Working Group III to the Fourth Assessment Report of the Intergovernmental Panel on Climate Change, Cambridge University Press, Cambridge.

Milbraith, L. W, 1996, *Learning to Think Environmentally While There is Still Time*, State University of New York Press, Albany, NY.

Miller, C 2004 (3 October), 'PM called talks to derail renewable energy', *The Age*, viewed 22 May 2011, <http://www.theage.com.au/articles/2004/10/02/1096527990014.html>.

Miller, D 2010, 'Get rid of centralised power to country areas', *The Age* (Melbourne), August 5, p. 16.

Milne, C 2008, 'Emissions trading: What does it mean?' *Green Tasmania*, Spring, pp. 15–16.

Milne, C 2009, (17 June), *Speech to the National Press Club*, unpublished.

Mitchell, A 2010, 'Democracy 101: the powerbrokers', *Sydney Morning Herald*, 4 July.

Mohr, SH & Evans, GM 2009, 'Forecasting coal production until 2100', *Fuel*, vol. 88, no. 11, pp. 2059–67.

MRET (Mandatory Renewable Energy Target) Review Panel 2003, *Renewable Opportunities: A Review of the Operation of the Renewable Energy (Electricity) Act 2000*, Mandatory Renewable Energy Target Review, viewed 22 May 2011, <http://www.mretreview.gov.au/report/index.html#summary>.

Mueller, D 1989, *Public Choice II: A Revised Edition of Public Choice*, Cambridge University Press, Cambridge.

Mulligan, M and Hill, S 2001, *Ecological Pioneers: A Social History of Australian Ecological Thought and Action*, Cambridge University Press, Cambridge.

Murray–Darling Basin Authority 2009, *Issues Paper: Development of Sustainable Diversion Limits for the Murray–Darling Basin*, MDBA publication no. 49/09, available for download at <http://www.mdba.gov.au/files/publications/sustainable-diversion-limits-issues-paper-12-11-09.pdf>.

Myors, P 2005 (with Rachel O'Leary & Rob Helstroom), 'Multi-unit residential building energy & peak demand study', *Energy News*, (NSW Department of Planning) vol. 23, no. 4.

NALWT (Northern Australia Land and Water Taskforce) 2009, *Sustainable Development of Northern Australia*, Northern Australia Land and Water Taskforce, Canberra.

NETT (National Emissions Trading Taskforce) 2006, *Possible Design for a National Greenhouse Gas Emissions Trading Scheme*, A Discussion paper prepared by the National Emissions Trading Taskforce, <http://www.caf.gov.au/ Documents/nett-discussion-paper.pdf>.

NLWRA (National Land and Water Resources Audit) 2001, *Australian Water Resources Assessment 2000*, NLWRA, Canberra.

NLWRA (National Land and Water Resources Audit) 2006, *Australian Water Resources Assessment 2006*, NLWRA, Canberra.

NOO (National Oceans Office) 2004, *South-East Regional Marine Plan: Implementing Australia's Oceans Policy in the South-east Marine Region*, National Oceans Office, Hobart.

NWC (National Water Commission) 2009, *Australian Water Reform 2009: Second Biennial Assessment of Progress in Implementation of National Water Initiative*, NWC, Canberra.

NWC (National Water Commission) 2010, *The Impacts of Water Trading in the Southern Murray–Darling Basin: An Economic, Social and Environmental Assessment*, NWC, Canberra.

NWC (National Water Commission) 2010, *Australian Environmental Water Management Report 2010*, NWC, Canberra.

Norby RJ, Warren JM, Iversen CM, Medlyn BE & McMurtrie RE 2010, 'CO2 enhancement of forest productivity constrained by limited nitrogen availability', *Proceedings of the National Academy of Sciences*, vol. 107, pp. 19368–19373.

NRMMC (Natural Resource Management Ministerial Council) 2006, *National Cooperative Approach to Integrated Coastal Zone Management: Framework and Implementation Plan*, Commonwealth of Australia, Canberra.

O'Connor, M 2008, 'A Short Environmental History of Australia', *The Edinburgh Review*, No. 128, February 2008, pp. 121 ff.

O'Connor, M 2009, 'On the replacement rate fallacy', web essay, <http://www.australianpoet.com/overloading.html#rrf>.

O'Connor, M 2011 (10 March), 'Fears for a big Australia', *The Drum*, Australian Broadcasting Corporation, <http://www.abc.net.au/unleashed/44896.html>.

O'Loughlin, L 1990, 'What the greenies lost in the election', *Chain Reaction*, vol. 60, pp. 18–20.

O'Sullivan, J 2010, 'Submission to population policy inquiry Local Government Association of Queensland', March 2010, p. 10, <http://www.lgaq.asn.au/c/document_library/get_file?p_l_id=621228&fileShortcutId=672191 >.

OECD 2007, *OECD Environmental Performance Reviews – Australia* viewed 12 July 2011 <http://www.oecd.org/dataoecd/29/54/40287648.pdf>.

Olsen, M 1965, *The Logic of Collective Action: Public Goods and the Theory of Groups*, Harvard University Press, Cambridge.

O'Neill, D 2004, *Review of Environmental Clearances for Permanent Interstate Water Trade*, Report to the Murray–Darling Basin Commission, Canberra.

Ophuls, W 1973, 'Leviathan or oblivion', in H. Daly (ed), *Toward a Steady State Economy*, WH Freeman, San Francisco.

Ostrom, E 1990, *Governing the Commons: the Evolution of Institutions for Collective Action*, Cambridge University Press, Cambridge.

Ostrom, E, Walker, J & Gardner, R 1992, 'Covenants with and without a sword: Self governance is possible', *The American Political Science Review*, vol. 86, no. 2, pp. 404–417.

Pagiola, S 2002, 'Paying for water services in Central America', in S Pagiola, J Bishop & N Landell Mills (eds), *Selling Forest Environmental Services: Market-based Mechanisms for Conservation and Development*, Earthscan, London, pp. 37–59.

Pagiola, S, Agostini, P, Gobbi, J, de Haan, C, Ibrahim, M, Murgueito, E, Ramirez, E, Rosales, M & Ruiz, J 2004, *Paying for Biodiversity Conservation Services in Agricultural Landscapes*, The World Bank Environment Department, Washington.

Pagiola, S, Bishop, J & Landell Mills N 2002, 'Market-based mechanisms for forest conservation and development', in S Pagiola, J Bishop & N Landell Mills (eds), *Selling Forest Environmental Services: Market-based Mechanisms for Conservation and Development*, London, Earthscan, pp. 1–14.

Parlane, L 2010, personal interview.

Parry, ML, Canziani, OF, Palutikof, JP, van der Linden, PJ & Hanson, CE (eds) 2007, *Climate Change 2007: Impacts, Adaptation and Vulnerability*, Contribution of Working Group II to the Fourth Assessment Report of the Intergovernmental Panel on Climate Change, Cambridge University Press, Cambridge.

Passmore, J 1975, *Man's Responsibility for Nature*, Duckworth, London.

Pearse, G 2007, *High and Dry: John Howard, Climate Change and the Selling of Australia's Future,* Penguin, Sydney.

Pearse, G 2009, 'Quarry Vision: Coal, climate change, and the end of the resources boom', *Quarterly Essay*, vol. 33, pp. 1–122.

Perrow, C 2007, *The Next Catastrophe: Reducing Our Vulnerabilities to Natural, Industrial and Terrorist Disasters,* Princeton University Press, Princeton NJ.

Peterson, D, Dwyer, G, Appels, D, & Fry, J 2004, *Modelling Water Trade in the Southern Murray–Darling Basin*, Productivity Commission, Melbourne.

Pike, D 1962, 'The smallholder's place in the Australian tradition', *Tasmanian Historical Research Association, Papers and Proceedings*, vol. 10, December, pp. 28–33.

Pittock, A Barrie 2009, *Climate Change: The Science, Impacts and Solutions*, CSIRO Publishing/Earthscan, Collingwood Australia, London UK.

Pittock, A Barrie 2010 (20 August), 'Climate change: the biggest campaign gaffe', *Sydney Morning Herald,* <http://www.smh.com.au/business/climate-change-the-biggest-campaign-gaffe-20100820-138hz.html>.

Powell, JM 1976, *Environmental Management in Australia, 1788-1914. Guardians, Improvers and Profit: an Introductory Survey*, Oxford University Press, Melbourne.

Powell, JM 1988, *An Historical Geography of Modern Australia: The Restive Fringe*, Cambridge University Press, Cambridge.

Prasser, S 2006, *Royal Commissions and Public Inquiries in Australia,* LexisNexis Butterworths, Chatswood, NSW.

Prasser, S 2010, 'Black Saturday inquiry puts the heat on Brumby', *The Australian,* August 2, p. 14.

Productivity Commission 2006, *Rural Water Use and the Environment: The Role of Market Mechanisms,* Productivity Commission, Melbourne.

Productivity Commission 2010, research paper *Population and Migration: Understanding the Numbers*, released 9 December 2010, <http://www.pc.gov.au/research/commissionresearch/population-migration>.

Puplick, C 2008, personal interview.

Pusey, M 1991, *Economic Rationalism in Canberra,* Cambridge University Press, Cambridge.

Pusey, M 2003, *The Experience of Middle Australia: The Dark Side of Economic Reform,* Cambridge University Press, Cambridge.

Pusey, M 2010, '25 years of neo-liberalism in Australia', in M Manne & D McNight (eds), *Goodbye to All That,* Black Inc., Melbourne, pp. 125–146.

Pyne, S J 2009, 'The human geography of fire: a research agenda', *Progress in Human Geography,* vol. 33, no. 4, pp. 443–446.

Pyne, SJ 1992, *Burning Bush: a Fire History of Australia*, Allen & Unwin, North Sydney.

Queensland Government 2009, *South East Queensland Climate Change Management Plan: Draft for Public Consultation*, DIP, Brisbane.

RAC (Resource Assessment Commission) 1992, *Forest and Timber Inquiry Final Report Overview*, Commonwealth of Australia, Canberra.

RAC (Resource Assessment Commission) 1993, *Coastal Zone Inquiry: Final Report*, Australian Government Publishing Service, Canberra.

Ratcliffe, F 1948, *Flying Fox and Drifting Sand*, Angus and Robertson, Sydney.

Rees, M, 2010 (25 July), 'Scientific horizons', *Reith Lecture*, radio program, ABC Radio National, Australian Broadcasting Corporation, Melbourne.

Ribaudo, M & Claassen, R 2006, 'Conservation policy overview', in K Weibe & N Gollehon (eds), *Agricultural Resources and Environmental Indicators*, United States Department of Agriculture, Washington, pp. 168–174.

Rittel, H & Webber, M 1973, 'Dilemmas in a general theory of planning', *Policy Sciences*, vol. 4, no. 2,pp. 155–167.

Roberts, G 2009, 'Why green leaders backed the carbon plan', *The Australian*, 9 May.

Roberts, S 1924, *History of Australian Land Settlement (1788-1920)*, Macmillan and Co., Melbourne.

Rolls, EC 1969, *They All Ran Wild*, Angus & Robertson, Sydney.

Rootes, C 2008, 'The first climate change election? The Australian general election of 24 November 2007', *Environmental Politics*, vol. 17, no. 3, pp. 473-480.

Rowe, M & McAllister, L 2006, 'The roles of commissions of inquiry in the policy process', *Public Policy and Administration*, vol. 21, no. 4, pp. 99–115.

Rudd, K 2009, 'Site clean-up program to pave the way for rebuilding', *Media release*, Office of the Prime Minister of Australia, 23 February, Canberra.

Saddler, H, Riedy, C & Passey, R 2004, *Geosequestration: What is it and how much can it contribute to a sustainable energy policy for Australia?* Discussion Paper No. 72, The Australia Institute, Canberra.

Sandercock, L 1976, *Cities for sale: property, politics and urban planning in Australia*, Heinemann, London.

Sanders, E 2010, 'Climate Action Summit 2010: A step forward', *Green Left Weekly*, 21 March, p. 1.

Schattschneider, EE 1960, *The Semisovereign People: A Realist's View of Democracy*, Holt, Reinehart and Winston, New York.

Schedvin, CB 1988, *Australia and the Great Depression*, Sydney University Press, Sydney.

Schlager, E & Ostrom E 1992, 'Property rights regimes and natural resources: A conceptual analysis', *Land Economics*, vol. 68, no. 3, pp. 249–262.

Schmitter, PC 1974, 'Still the century of corporatism?', *Review of Politics*, vol. 36, pp.85–133.

Schofield, J 8 June 2010, 'The mining tax ads counter the miners' fear campaign', *The Punch*, (News Ltd), viewed 28 March 2011, <http://www.thepunch.com.au/articles/the-mining-tax-ads-counter-the-miners-fear-campaign/>.

Scott, S 2011, 'Bigger Australia a must: Immigration', *Australian Financial Review*, 5 January 2011.

Seddon, G 1997, *Landprints: Reflections on Place and Landscape*, Cambridge University Press, Cambridge.

Senate Standing Committee on Rural and Regional Affairs and Transport 2008, *Water Management in the Coorong and the Lower Lakes*, Parliament of Australia.

SEQWater 2002, *SEQWater's Dams*, SEQWater Corporation, Brisbane.

Shadwick, M 2002, *A Viable and Sustainable Water Industry: National Competition Council staff discussion paper*, Ausinfo, Canberra.

Shanahan D 2009, 'Blunderful Copenhagen kills the ETS early poll', *The Australian*, 18 December, viewed 4 July 2011, <http://www.theaustralian.com.au/news/opinion/blunderful-copenhagen-kills-ets-early-poll/story-e6frg6zo-1225811525228>.

Shanahan, D 2007, 'Taskforce looks to clear path for food bowl's shift north', *The Australian*, Sydney.

Shaw, A 1990, 'Colonial settlement 1788–1945', in D Williams (ed), *Agriculture in the Australian Economy*, Sydney University Press, Sydney, pp. 1–18.

Silvester, J 2010, 'Hundreds in bushfire class action', *The Age* (Melbourne), June 19, p.1.

Sinclair, P 2001, *The Murray: A River and its People*, Melbourne University Press, Carlton.

SKM (Sinclair Knight Merz) CSIRO & BRS (The Bureau of Rural Sciences) 2010, *Surface and/or Groundwater Interception Activities: Initial Estimates, Waterlines Report*, National Water Commission, Canberra.

SMH 2006,'Revealed the Sydney Flats Squeeze', *Sydney Morning Herald*, 26 December 2006, p. 1; cf. ibid, 27 December 2006 'Residents fear impact of imposed population explosion'.

Smith, D 1990, *Continent in Crisis*, Penguin, Ringwood.

Smith, D 1994, *Saving a Continent: Towards a Sustainable Future*, UNSW Press, Kensington.

Smith, D 2003, 'Water resources management', in S Dovers & S Wild River (eds), *Managing Australia's Environment*, The Federation Press, Sydney.

Smith, GJ 1990, *Toxic Cities and the Fight to Save the Kurnell Peninsula*, UNSW Press, Kensington.

Smyth, C 2010, (Australian Conservation Foundation Healthy Oceans Campaigner) personal communication.

Sobels, J, Richardson, S, Turner, G, Maude, A, Tan, Y, Beer, A, Wei, Z 2010, (July) *Research into the long-term physical implications of Net Overseas Migration*, Department of Immigration and Citizenship, viewed 4 June 2011, <http://www.immi.gov.au/media/publications/research/ pdf/physical-implications-migration-fullreport.pdf>.

Solomon, S, Qin, D, Manning, M, Chen, Z, Marquis, M, Avery, KB, Tignor, M. & Miller, HL (eds) 2007, *Climate Change 2007: The Physical Science Basis*, Contribution of Working Group I to the Fourth Assessment Report of the Intergovernmental Panel on Climate Change, Cambridge University Press, Cambridge.

Sørensen, B 2008, 'A renewable energy and hydrogen scenario for northern Europe', *International Journal of Energy Research*, vol. 32, no. 5, pp. 471–500.

Sorenson, J 1997, 'National and international efforts at integrated coastal management: Definitions, achievements and lessons', *Coastal Management*, vol. 25, no. 1, pp. 3–41.

Sørensen, B & Meibom, P 2000, 'A global renewable energy scenario', *International Journal of Global Energy Issues*, vol. 13, nos 1/2/3, doi: 10.1504/IJGEI.2000.000869.

SPPA (Stable Population Party of Australia) 2010, *Population Policy*, Stable Population Party of Australia, online at <http://www.populationparty.com/>.

Staples, J 2006, 'NGOs out in the cold: The Howard government policy towards NGOs', Democratic Audit of Australia, viewed 30 December 2006, <http://www.democratic.audit.anu.edu.au>.

Staples, J, 2008, 'Attacks on NGO "accountability': Questions of governance or the logic of public choice theory?', in J. Barraket (ed), *Strategic Issues for the Not-for-Profit Sector*, UNSW Press, Sydney.

Staples, J 2009, 'Environment: "What was right was popular"', in Bloustein, G, Coomber, B & Mackinnon, A, (eds), *The Hawke Legacy*, Wakefield, Adelaide.

State of the Environment Advisory Council 1996, *State of the Environment Australia 1996,* CSIRO Publishing, Collingwood Vic.

Steketee, M 2008, 'Overcompensation blunts ETS's value', *The Australian* 18 December accessed 18.12.08 <http://www.theaustralian.news.com.au/story/0,25197, 24815450-7583,00.html>.

Steketee, M 2010, 'Labor's great climate change choke', *The Australian*, 31 July, viewed 4 July 2011, <http://www.theaustralian.com.au/national-affairs/labors-great-climate-change-choke/story-fn59niix-1225898997863>.

Stern, N 2006, *Stern Review: The economics of climate change,* October, HM Treasury, viewed 22 May 2011, <http://www.hm-treasury.gov.uk/sternreview_index.htm>.

Stewart, C 2010, 'Fires challenge the tree-change dream', *The Australian,* August 2, p. 13.

Still, J 2010, 'Overpopulate and perish, says Anglican body', 12 May 2010, web-article, <http://www.melbourne.anglican.com.au/NewsAndViews/Pages/Overpopulate-and-perish,-says-Anglican-body.aspx>.

Stilwell, F 2009, (22 February), *Economists Speak Out Against Flawed Carbon Trading Scheme,* The Evatt Foundation, <http://evatt.org.au/news/498.html>.

Stoneham, G, Chaudry, V, Ha, A & Strappazon, L 2003, 'Auctions for conservation contracts: An empirical examination of Victoria's bush tender trial', *Australian Journal of Agricultural and Resources Economics*, vol. 47, no. 4, pp. 477–500.

Sturgess, G & Wright, M (1993), *Water Rights in Rural New South Wales: The Evolution of a Property Rights System*, Centre for Independent Studies, St Leonards.

Stutchbury, M 2008, 'Kyoto Kev's still a green hero when you get into the details', *The Australian* 19th December, p. 28.

Sustainable Population Australia 2011, *Home Page for Sustainable Population Australia,* Sustainable Population Inc. Weston Creek ACT viewed 30 June 2011, <www.population.org.au>.

Swiss Re 2010, *Natural Catastrophes and Man-Made Disasters in 2009, Sigma Report,* 1/2010, Swiss Re, Zurich.

Tabara, D 2005, *HarmoniCOP Integration report – Sustainability learning for river basin management and planning in Europe,* Work Package 6, August, European Commission.

Taylor, G 1943, *Australia: A Study of Warm Environments and Their Effect on British Settlement,* 2nd edn, Methuen, London.

Taylor, P 1994, *Growing Up: Forestry in Queensland,* Allen & Unwin, Brisbane.

Teague, B, McLeod, R & Pascoe, S 2009, *2009 Victorian Bushfires Royal Commission,* Interim Report, Government Printer, Melbourne.

Teague, B, McLeod, R & Pascoe, S 2010, *2009 Victorian Bushfires Royal Commission,* Final Report, Government Printer, Melbourne. (See also Victorian Bushfires Royal Commission 2010.)

Teske, S & Vincent, J 2008, *Energy [R]evolution: A Sustainable Australia Energy Outlook,* Greenpeace Australia Pacific and European Renewable Energy Council, viewed 22 May 2011, <http://www.energyblueprint.info>.

The Australian, 2007, *The Australian,* December 11, p. 1.

Thom, BG & Harvey, N 2000, 'Trigger for late twentieth century reform of Australian coastal management', *Australian Geographical Studies,* vol. 38, no. 3, pp. 275–290.

Thomas, I 2007, *Environmental Policy: Australian Practice in the Context of Theory,* Federation Press, Annandale, NSW.

Thompson, E 1980, 'The 'Washminster' Mutation', in P Weller and D Jaensch (eds), *Responsible Government in Australia,* Drummond & the Australasian Political Studies Association, Victoria, pp. 32–40.

Thomson, K 2009a, Speech in Federal Parliament by Hon. Kelvin Thomson, 17 August 2009, <.aph.gov.au//reps/dailys/dr170809.pdf>.

Thomson, K 2009b, 'There is an alternative to runaway population: Kelvin Thomson's 14-point plan for population reform', <http://www.kelvinthomson.com.au/Editor/assets/speeches_2009/091111 percent20population percent20reform percent20paper percent20ac.pdf>.

Tilzey, M 2006, 'Neo-liberalism, the WTO and new modes of agri-environmental governance in the European Union, the USA and Australia ', *International Journal of Sociology of Food and Agriculture,* vol. 14, no. 1, pp. 1-28.

Toyne, P & Balderstone, S 2003, 'The Environment', in S Ryan and T Bramston (eds), *The Hawke Government: A Critical Retrospective,* Pluto Press, North Melbourne.

Toyne, P 1994, *The Reluctant Nation: Environment, Politics & the Law in Australia,* ABC Books, Sydney.

Toyne, P 2008, personal interview.

Tupper, G 2008, personal interview.

UNEP (United Nations Environment Program) 2007, *Global Environmental Outlook 4,* UNEP, Nairobi.

UNEP (United Nations Environment Program) 2007, 'Kick the habit, a UN guide to climate neutrality', Chapter 4, UNEP ebook, <http://www.unep.org/publications/ebooks/kick-the-habit/pdfs/4.Actors_en.pdf>.

Van Onselen, P 2010, 'Politics trumps a moral challenge' *The Australian*, 29 April, viewed 4 July 2011, <http://www.theaustralian.com.au/news/features/politics-trumps-a-moral-challenge/story-e6frg6z6-1225859592923>.

VCC (Victorian Coastal Council) 2008, *Victorian Coastal Strategy*, Victorian Coastal Council. Melbourne.

Victorian Bushfires Royal Commission (see Teague *et al.*).

Victorian Liberal Nationals Coalition 2010, *Research Report: 2009 Victorian Royal Commission Interim Report Recommendations Still Not Implemented by the Brumby Government*, Melbourne.

Vlek, C 2010, 'Judicious management of uncertain risks: I. Developments and criticisms of risk analysis and precautionary reasoning', *Journal of Risk Research*, vol. 13, no. 4, pp. 517–543.

Wahlquist, A 2008, *Thirsty Country: Options for Australia*, Allen & Unwin, Crows Nest.

Walker, KJ 1994, *The Political Economy of Environmental Policy: An Australian Introduction*, University of New South Wales Press, Sydney.

Walker, KJ (ed) 1992, *Australian Environmental Policy*, UNSW Press, Kensington.

Walker, KJ 1989, 'The state in environmental management: The ecological dimension', *Political Studies*, vol. 37, March, pp. 25–39.

Walker, KJ 1994, *The Political Economy of Environmental Policy: An Australian Introduction*, UNSW Press, Kensington.

Walker, KJ 1995, 'Environmental policy: Adequacy of government response', unpublished paper delivered at the Australasian Political Studies Conference, Melbourne, September.

Walker, KJ & Crowley, K 1999, *Australian Environmental Policy: Studies in Decline & Devolution*, UNSWPress, Kensington, NSW.

Walker, R 1993, 'Evaluating the financial performance of Australian water authorities', in M Johnson & S Rix (eds), *Water in Australia, Managing Economic, Environmental and Community Reform*, Pluto Press, Leichhardt.

Ward, R 1975, *Australia: A Short History*, Ure Smith, Sydney.

Warhurst, J 2004, 'Patterns and directions in Australian politics over the past fifty years', *Australian Journal of Politics and History*, vol. 50, no. 2, pp.163–177.

Waterson, D 1968, *Squatter, Selector and Storekeeper*, Sydney University Press, Sydney.

Watson, C 2008, 'Boundless Plains', maps of Australia by climate, rainfall and soil type, <http://www.australianpoet.com/boundless.html>.

Webb, L, Whitelock, D & J Le Gay Brereton (eds) 1969, *The Last of the Lands*, Jacaranda Press Pty Ltd, Milton, Qld.

Weir, J 2009, *Murray River Country: An ecological dialogue with traditional owners*, Aboriginal Studies Press, Canberra.

Wescott, G 2000, 'Development and initial implementation of Australia's "integrated and comprehensive" Oceans Policy', *Ocean and Coastal Management*, vol. 43, no. 10-11, pp. 853–878.

Wescott, G 2009 'Stimulating vertical integration in coastal management in a federated nation: the case of Australian coastal policy reform', *Coastal Management*, vol. 37, no. 6, pp. 501–513.

Wescott, G & Fitzsimons, J 2011, 'Stakeholder involvement and interplay in coastal zone management and marine protected area planning', Chapter 17, in W Gullett, J Vince, and C Schofield (eds), *Marine Management in Australia*, LexisNexis Butterworths, Chatswood, NSW, pp. 225–238.

Whitelock, D 1971, *A Dirty Story: Pollution in Australia*, Sun Books, Melbourne.

Wilby, R & Gell, P 1994, 'The impact of forest harvesting on water yield: modelling hydrological changes detected by pollen analysis', *Journal of Hydrological Sciences*, vol. 39, no. 5, pp. 471–486.

Wilkie, A 2004, *Axis of Deceit*, Black Inc., Melbourne.

Wilson, G 2001, 'From productivism to post-productivism ... and back again? Exploring the (un)changed natural and mental landscapes of European agriculture', *Transactions of the Institute of British Geographers*, vol. 26, no. 1, pp. 7–102.

Wilson, J 2010, 'Remember the bushfires' other victims', *The Age*, January 1–2.

Wilson, L & Rintoul, S 2010, 'Brumby delays on report', *The Australian*, August 3, p. 4.

Wong, P 2008, Rudd Government to Invest $12.9 billion in Water, *Media release no. PW 56/08*, viewed 13 July 2011 <http://www.climatechange.gov.au/~/media/Files/minister/previous%20minister/wong/2008/Media%20Releases/April/mr20080429.pdf >.

Wood, M, Stephens, N, Allison, B & Howell, C 2001, *Plantations of Australia*, Bureau of Rural Sciences, Canberra.

Wootten, H 1987, 'Elections and environmental politics: The search for consensus', *Habitat*, vol. 15, pp. 7–9.

Wright, M & Hearps, P 2010, *Australian Sustainable Energy: Zero Carbon Australia Stationary Energy Plan*, University of Melbourne Energy Research Institute and Beyond Zero Emissions, Melbourne.

WWF International, (with Ecofys & OMA) 2011, *The Energy Report: 100% renewable energy by 2050*, viewed 19 May 2011, <http://wwf.panda.org/what_we_do/footprint/climate_carbon_energy/energy_solutions/renewable_energy/sustainable_energy_report>.

Yates, A & Bergin, A 2009, *Hardening Australia: Climate Change and National Disaster Resilience*, Special Report, August, Australian Strategic Policy Institute, Canberra.

Yencken, D & Wilkinson, D 2000, *Resetting the Compass: Australia's Journey Towards Sustainability*, CSIRO Publishing, Collingwood, Vic.

Youl, R, Marriott, S, & Nabben, T 2006, Landcare in Australia, SILC & Rob Youl Consulting Pty Ltd, Melbourne.

Index

G

H

I

K

L

M

N

O

P

Q

R

S

T

U

V

W

Z